AF379548

Advances in X-Ray Spectroscopy of Laser Plasmas

Eugene Oks
Physics Department, Auburn University, Auburn, Alabama, USA

IOP Publishing, Bristol, UK

© IOP Publishing Ltd 2020

All rights reserved. No part of this publication may be reproduced, stored in a retrieval system or transmitted in any form or by any means, electronic, mechanical, photocopying, recording or otherwise, without the prior permission of the publisher, or as expressly permitted by law or under terms agreed with the appropriate rights organization. Multiple copying is permitted in accordance with the terms of licences issued by the Copyright Licensing Agency, the Copyright Clearance Centre and other reproduction rights organizations.

Permission to make use of IOP Publishing content other than as set out above may be sought at permissions@ioppublishing.org.

Eugene Oks has asserted his right to be identified as the author of this work in accordance with sections 77 and 78 of the Copyright, Designs and Patents Act 1988.

ISBN   978-0-7503-3375-7 (ebook)
ISBN   978-0-7503-3373-3 (print)
ISBN   978-0-7503-3376-4 (myPrint)
ISBN   978-0-7503-3374-0 (mobi)

DOI   10.1088/978-0-7503-3375-7

Version: 20200701

IOP ebooks

British Library Cataloguing-in-Publication Data: A catalogue record for this book is available from the British Library.

Published by IOP Publishing, wholly owned by The Institute of Physics, London

IOP Publishing, Temple Circus, Temple Way, Bristol, BS1 6HG, UK

US Office: IOP Publishing, Inc., 190 North Independence Mall West, Suite 601, Philadelphia, PA 19106, USA

In memory of recently passed away

A Ya Faenov

with appreciation of his truly outstanding contribution

to the experimental spectroscopy of laser-produced plasmas.

Contents

Author biography

Eugene Oks

 **Eugene Oks** received his PhD degree from the Moscow Institute of Physics and Technology, and later the highest degree of Doctor of Sciences from the Institute of General Physics of the Academy of Sciences of the USSR by the decision of the Scientific Council led by the Nobel Prize winner, academician A M Prokhorov. According to the Statute of the Doctor of Sciences degree, this highest degree is awarded only to the most outstanding PhD scientists who founded a new research field of a great interest. Oks worked in Moscow (USSR) as the head of a research unit at the Center for Studying Surfaces and Vacuum, then—at the Ruhr University in Bochum (Germany) as an invited professor, and for the last 30 years— at the Physics Department of the Auburn University (USA) in the position of Professor. He conducted research in five areas: atomic and molecular physics; plasma physics, laser physics; nonlinear dynamics; and astrophysics. He founded/co-founded and developed new research fields, such as intra-Stark spectroscopy (a new class of nonlinear optical phenomena in plasmas), masing without inversion (advanced schemes for generating/amplifying coherent microwave radiation), and quantum chaos (nonlinear dynamics in the microscopic world). He also developed a large number of advanced spectroscopic methods for diagnosing various laboratory and astrophysical plasmas—the methods that were then used and are used by many experimental groups around the world. He has published about 450 papers and 7 books, including the books *Plasma Spectroscopy: The Influence of Microwave and Laser Fields*, *Stark Broadening of Hydrogen and Hydrogenlike Spectral Lines in Plasmas: The Physical Insight*, *Breaking Paradigms in Atomic and Molecular Physics*, *Diagnostics of Laboratory and Astrophysical Plasmas Using Spectral Lineshapes of One-, Two, and Three-Electron Systems*, *Unexpected Similarities of the Universe with Atomic and Molecular Systems: What a Beautiful World*, and *Analytical Advances in Quantum and Celestial Mechanics: Separating Rapid and Slow Subsystems*. He is the Chief Editor of the journal *International Review of Atomic and Molecular Physics*. He is a member of the Editorial Boards of two other journals: the *Open Journal of Microphysics* and the journal *Open Physics*. He is also a member of the International Program Committees of the two series of conferences: Spectral Line Shapes, as well as Zvenigorod Conference on Plasma Physics and Controlled Fusion.

IOP Publishing

Advances in X-Ray Spectroscopy of Laser Plasmas

Eugene Oks

Chapter 1

Introduction

This book presents advances in x-ray spectroscopy of:
- plasmas interacting with a laser radiation;
- laser-induced plasmas.

We focus mostly (but not exclusively) in advances in x-ray spectroscopic diagnostics. These advances are mainly due to the progress in theoretical and experimental studies of the shapes of x-ray spectral lines. The primary reason is that the spectral line shapes are practically independent of the choice of a particular model of the plasma state. This is a clear distinction from other diagnostics that depend on the choice of the plasma state, which could result in a large uncertainty.

The first theoretical underpinnings for analyzing x-ray spectra for plasma diagnostics were developed for astrophysical purposes. It is considered that the beginning of the observational x-ray astrophysics dates to the year 1962 where x-ray binary source Scorpius X-1 was discovered [1]. In the intervening dozens of years, tremendous progress has been made in observational x-ray astrophysics—see, for example, review [2] and references therein. X-ray spectroscopy enables studying a wide variety of astrophysical objects. In particular, diagnostics based on x-ray spectral lines are used, for instance, for measuring densities, temperatures, ionization balance, and abundances—see, for example, review [3] and references therein.

These techniques were further developed and used for diagnostics of various laboratory plasmas. X-ray spectroscopy is a set of multi-parameter methods for plasma diagnostics. In its traditional form it can provide information, for example, about the electron density, the temperature of electrons and ions, the densities of radiating atoms and ions, and the ionization balance—see, for instance, review [4], as well as book [5] and references therein.

The most fruitful results in the traditional x-ray diagnostics of laser plasmas were achieved due to advances in the theory of the Stark broadening of ion spectral lines in plasmas—see, for example books [6–9]. A brief overview of some of these

doi:10.1088/978-0-7503-3375-7ch1

© IOP Publishing Ltd 2020

theoretical advances is presented in appendix A of the current book. These advances provided opportunities for relatively accurate measurements of the electron density, as well as of the temperature of electrons and ions.

More recent advances in theoretical and experimental laser plasma diagnostics based on the x-ray spectral line shapes have significantly expanded the scope of measured parameters—see, for instance, book [9] and references therein. It became possible to study the development of Langmuir waves, ion acoustic waves, and transverse electromagnetic waves induced by various laser–plasma interactions. Also, these advances enabled the possibility to study other nonlinear processes caused by laser–plasma interactions, such as, for instance, parametric decay instabilities. Last, but not least: it has also become feasible to obtain information about the rates of charge exchange between multicharged ions—information virtually inaccessible for other methods. Details on all of these advances can be found in papers [10–35] (listed in chronological order) and are presented in the current book.

Additional important information can be obtained through the polarization analysis of x-ray spectral line shapes. These methods, presented in papers [36–38] were developed for powerful Z-pinches, but they are applicable also for laser plasmas.

The practical importance of this entire research area is the following. First, it is indispensable for one of the two major directions in the quest for controlled nuclear fusion—namely, for laser fusion. Second, it is an important tool in studying matter under the extreme conditions produced by super-high-intensity lasers. Third, it can provide atomic reference data (such as the rate of charge exchange between multicharged ions) virtually inaccessible by other methods. Fourth, it shows the way to creating plasma-based tunable x-ray lasers. Fifth, it opens up new avenues for laboratory modeling of physical processes in astrophysical objects and a better understanding of intense laser–plasma interactions.

This book is structured as follows. Chapter 2 presents theoretical and experimental studies of charge-exchange-caused dips (X-dips) in profiles of x-ray spectral lines. It also presents the application of the X-dip phenomenon to the experimental determination of rate coefficients of charge exchange between multicharged ions in plasmas.

Chapter 3 is devoted to diagnostics of non-relativistic laser–plasma interactions based on their effects on x-ray spectral line shapes. Most of the attention is given to the phenomenon of Langmuir-wave-caused dips (L-dips) and their applications to measuring parameters of plasmas and of the laser- or laser-caused-fields in plasmas.

Chapter 4 presents theoretical and experimental studies of relativistic laser–plasma interactions based on their effects on x-ray spectral line shapes. The corresponding diagnostics provide, in particular, experimental information about various nonlinear processes in plasmas, such as, for example, the parametric instabilities.

In chapter 5 the discussion is focused on the possibilities of spectroscopic measurements of GigaGauss or multi-GigaGauss magnetic fields, developing during interactions of plasmas with super-intense laser radiation. These possibilities have employed effects of these ultra-intense magnetic fields on the locations and width of L-dips in x-ray spectral line profiles.

Chapter 6 presents concluding remarks. It includes a brief description of several works that did not fit in the scope of chapters 2–5.

The appendices provide additional details on the topics presented in chapters 2–6.

References

[1] Giacconi R, Gursky H, Paolini F and Rossi B B 1962 *Phys. Rev. Lett.* **9** 439

[2] Pounds K 2002 *Philos. Trans. R. Soc. Lond.* A **360** 1905

[3] Kahn S M, Behar E, Kinkhabwala A and Savin D W 2002 *Philos. Trans. R. Soc. Lond.* A **360** 1923

[4] Giulietti D and Gizzi L A 1998 *La Riv. del Nuovo Cimento* **21** 1

[5] Kunze H-J 2009 *Introduction to Plasma Spectroscopy* (Heidelberg: Springer)

[6] Griem H R 1997 *Principles of Plasma Spectroscopy* (Cambridge: Cambridge University Press)

[7] Fujimoto T 2004 *Plasma Spectroscopy* (Oxford: Clarendon)

[8] Oks E 2006 *Stark Broadening of Hydrogen and Hydrogenlike Spectral Lines in Plasmas: The Physical Insight* (Oxford: Alpha Science International)

[9] Oks E 2017 *Diagnostics of Laboratory and Astrophysical Plasmas Using Spectral Lineshapes of One-, Two-, and Three-Electron Systems* (Hackensack, NJ: World Scientific)

[10] Oks E and Leboucher-Dalimier E 2000 *Phys. Rev.* E **62** R3067

[11] Oks E and Leboucher-Dalimier E 2000 *J. Phys. B: At. Mol. Opt. Phys.* **33** 3795

[12] Leboucher-Dalimier E, Oks E, Dufour E, Sauvan P, Angelo P, Schott R and Poquerusse A 2001 *Phys. Rev.* E **64** 065401

[13] Leboucher-Dalimier E, Oks E, Dufour E, Angelo P, Sauvan P, Schott R and Poquerusse A 2002 *Eur. Phys. J.* D **20** 269

[14] Belyaev V S *et al* 2004 *J. Exp. Theor. Phys.* **99** 708

[15] Renner O, Dalimier E, Oks E, Krasniqi F, Dufour E, Schott R and Foerster E 2006 *J. Quant. Spectrosc. Radiat. Transf.* **99** 439

[16] Gavrilenko V P *et al* 2006 *J. Phys. A: Math. Gen.* **39** 4353

[17] Dalimier E, Oks E, Renner O and Schott R 2007 *J. Phys. B: At. Mol. Opt. Phys.* **2007** 909

[18] Sauvan P, Dalimier E, Oks E, Renner O, Weber S and Riconda C 2009 *J. Phys. B: At. Mol. Opt. Phys.* **42** 195001

[19] Renner O *et al* 2009 *High Energy Density Phys.* **5** 139

[20] Sauvan P, Dalimier E, Oks E, Renner O, Weber S and Riconda C 2010 *Int. Rev. At. Mol. Phys.* **1** 123

[21] Oks E and Dalimier E 2011 *Int. Rev. At. Mol. Phys.* **2** 43

[22] Sauvan P, Faenov A Y, Dalimier E, Oks E, Pikuz T A and Skobelev I Y 2011 *Int. Rev. At. Mol. Phys.* **2** 93

[23] Renner O, Dalimier E, Liska R, Oks E and Šmíd M 2012 *J. Phys. Conf. Ser.* **397** 012017

[24] Dalimier E and Oks E 2012 *Int. Rev. At. Mol. Phys.* **3** 85

[25] Oks E *et al* 2014 *J. Phys. B: At. Mol. Opt. Phys.* **47** 221001

[26] Oks E, Dalimier E, Faenov A and Renner O 2014 *J. Phys. Conf. Ser.* **548** 012030

[27] Dalimier E, Oks E and Renner O 2014 *Atoms* **2** 178

[28] Oks E *et al* 2015 *Opt. Express* **23** 31991

[29] Faenov A Y *et al* 2016 *Quantum Electron.* **46** 338

[30] Dalimier E *et al* 2017 *J. Phys. Conf. Ser.* **810** 012004
[31] Dalimier E and Oks E 2017 *J. Phys. B: At. Mol. Opt. Phys.* **50** 025701
[32] Oks E *et al* 2017 *Opt. Express* **25** 1958
[33] Oks E *et al* 2017 *J. Phys. B: At. Mol. Opt. Phys.* **50** 245006
[34] Dalimier E and Oks E 2018 *Atoms* **6** 60
[35] Oks E, Dalimier E and Angelo P 2019 *Spectrochim. Acta* B **157** 1
[36] Demura A V and Oks E 1998 *IEEE Trans. Plasma Sci.* **26** 1251
[37] Clothiaux E J, Oks E, Weinheimer J, Svidzinski V and Schulz A 1997 *J. Quant. Spectrosc. Radiat. Transf.* **58** 531
[38] Weinheimer J, Oks E, Clothiaux E J, Schulz A and Svidzinski V 1998 *IEEE Trans. Plasma Sci.* **26** 1239

IOP Publishing

Advances in X-Ray Spectroscopy of Laser Plasmas

Eugene Oks

Chapter 2

Charge-exchange-caused dips in spectral lines emitted by laser-produced plasmas

2.1 Overview of the theory of the charge-exchange-caused dips in profiles of hydrogenic spectral lines

Local depressions in spectral line profiles, caused by charge exchange (CE) in plasmas having ions of at least two different nuclear charges Z and $Z' \neq Z$, are called X-dips. It should be emphasized that CE is one of the most fundamental processes in nature and it has a profound practical significance (see, e.g. the book [1]). For example, CE between multicharged impurity ions and hydrogen (or deuterium, or tritium) atoms in magnetic fusion devices (e.g. tokamaks) affects the feasibility of controlled fusion—see, e.g. [2, 3] and references therein. CE in magnetic fusion devices is also employed for measuring the density of impurity ions (see, e.g. [1]). CE also serves as an effective mechanism for achieving population inversion x-ray lasers [1, 4–11]. In addition, CE controls the functioning of ion storage devices (see, e.g. [12]). Finally, CE is significant for the solar plasma and for planetary nebulae [1].

Let us say upfront that theoretical and experimental studies of X-dips are important not only from the fundamental point of view, but also from the practical point of view. These studies provide a way for the experimental determinations of the rates of CE between multicharged ions—information virtually impossible to get by other experimental means.

The X-dip was first observed in 1995 in the profile of the neutral hydrogen line H_α emitted from a helium plasma of the gas-liner pinch [13]. It was found in the blue wing of the H-alpha line hydrogen atoms ($Z = 1$) perturbed by fully stripped helium ($Z' = 2$)—for a limited range of electron densities around 10^{18} cm^{-3} [13]. Paper [13] also contained the first draft of the underlying theory. It focused on the fact that CE provides an additional channel for the decay of the excited state of the radiating ion, from which the spectral line originates, and thus shortens the lifetime of this state.

© IOP Publishing Ltd 2020

Namely, CE takes place at the vicinity of avoided crossings of terms of the quasimolecule ZeZ', made up of a H-like radiating ion Z and a perturbing fully stripped ion Z'. Consequently, at an avoided crossing there occurs an additional dynamical broadening γ_{CE} of the spectral line—additional to the non-CE dynamical broadening mechanisms yielding γ_{nonCE}. Because this additional broadening is effective only in a small vicinity δR of the internuclear distance R, corresponding to the avoided crossing, a local feature can show up in the corresponding place of the line profile—namely, the X-dip having generally the bump–dip–bump structure.

In the intervening years, further developments of the X-dip theory revealed the following picture. Let us start from the fact that it was well-known that the shape of spectral lines of a radiating atom/ion (radiator) in a gas or in a plasma depends on the energy terms of the combined quantum system 'radiator + perturber(s)'. Sometimes the energy difference between the terms involved in the radiative transition, being plotted versus the radiator–perturber separation, demonstrated extrema. This situation has been studied for over 30 years—both theoretically and experimentally (see, e.g. [14–24] and references therein). The paradigm based on these studies was that the extrema in the transition energy result in *satellites* in spectral line profiles [14–24].

However, in paper [25] it was demonstrated that the extrema in the transition energy can also result in *dips* in spectral line profiles. It was shown that for the practically important case where the extremum in the transition energy is due to the avoided crossing (where CE occurs), its spectral signature most probably should be a dip (X-dip) rather than a satellite. Here are some details.

We consider a radiative transition between two terms corresponding to some Stark component. We use atomic units and therefore employ the same notation $f(R)$ for both the transition energy and the transition frequency; R is the distance between the radiator and the perturbing atom or ion. We denote as $g(R)$ the area-normalized probability distribution of the quantity R. In the quasistatic approximation, the area-normalized profile $I(\Delta\omega)$ of the Stark component versus the detuning $\Delta\omega$ from the unperturbed frequency ω_0 is usually given by

$$I(\Delta\omega) = \int_0^\infty dR\ G(R)\delta[\Delta\omega - f(R)], \qquad G(R) = g(R)J(R)/J(\infty), \qquad (2.1)$$

where $J(R)$ is a frequency-integrated relative intensity of the Stark component.

We consider a vicinity of some particular distance R_0 corresponding to a small part of the component profile around $\Delta\omega_0 = f(R_0)$. In a relatively simple case where $f(R)$ does not have an extremum at $R = R_0$, from equation (2.1) one usually obtains

$$I(\Delta\omega_0) = G(R_0)/|f'(R_0)|. \qquad (2.2)$$

However, if $f(R)$ has an extremum at $R = R_0$, so that the first derivative $f'(R)$ vanishes, then equation (2.2) as well as equation (2.1) at $\Delta\omega = \Delta\omega_0$ becomes inapplicable. Physically this means that some feature in the profile may arise in the vicinity of $\Delta\omega_0$. For obtaining a finite value of $I(\Delta\omega_0)$, one should allow for additional broadening mechanisms and substitute the delta-function in equation (2.1) by a more realistic profile, such as a Lorentzian or a Gaussian. For example, for the Lorentzian having a full-width at half-maximum (FWHM) γ, as shown in section 3.3.2 of the book [26], the intensity $I(\Delta\omega_0)$ becomes:

$$I(\Delta\omega_0) \approx G(R_0)\int_{-\infty}^{\infty} d(R - R_0)[\gamma/(2\pi)]/\{(\gamma/2)^2 + [f''(R_0)/2]^2(R - R_0)^4\}$$
$$= G(R_0)\{2/[\gamma|f''(R_0)|]\}^{1/2}. \tag{2.3}$$

The above result (2.3) was obtained under the assumption that the function $G(R)$ is the slowest out of two factors in the integrand. The justification of this assumption was provided in paper [25].

We are interested primarily in the situation where an extremum in the energy difference between an upper term a and a lower term a_0 is caused by an avoided crossing of the radiator's term a with some perturber's term a'. We denote as $f_-(R)$ the transition frequency between the *original* terms a and a_0 at $R \leqslant R_0$ and as $f_+(R)$ the transition frequency between the *original* terms a' and a_0 at $R \geqslant R_0$. Here by *'original'* we mean the terms as they would be if the terms a and a' were not coupled and therefore crossed (figure 2.1).

In the vicinity of $R = R_0$, as a result of the avoided crossing, there occurs a transition of the energy difference from $f_-(R)$ to $f_+(R_0)$ as well as a transition of the slope from $f_-'(R_0)$ to $f_+'(R_0)$. If $f_-'(R_0)$ and $f_+'(R_0)$ have opposite signs, the avoided crossing causes an extremum in the energy/frequency difference (figure 2.1).

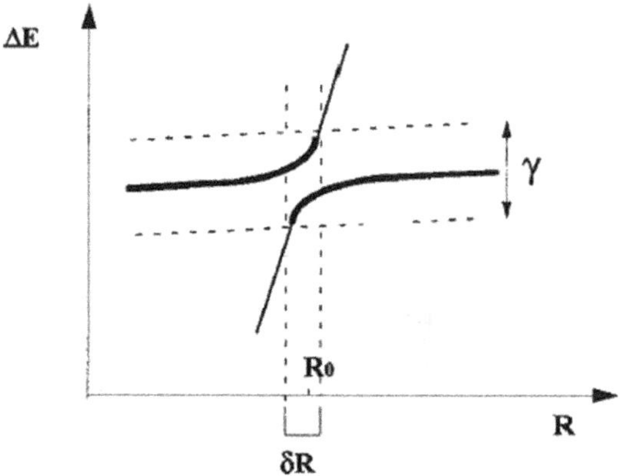

Figure 2.1. Transition energies $f_-(R) = E_a - E_{a0}$ and $f_+(R) = E_{a'} - E_{a0}$ versus the radiator–perturber separation R, plotted in a vicinity δR of an avoided crossing of the perturber's term a' with the radiator's term a at $R = R_0$ in the course of the radiative transition from the term a to the term a_0 [25]. The transition energy $f_-(R)$ actually occupies a band of a width γ (shown by dashed lines) controlled primarily by the dynamical broadening caused by electron and ion microfields in a plasma. The radiator's transition energy modified by the avoided crossing is shown by the bold line. In the interval δR, the transition energy has two branches, corresponding to the fact that the wave function of the radiator's term in this interval is a linear combination of wave functions of two different energies.

Combining equations (2.2) and (2.3), it is easy to calculate the ratio of the true intensity $I(\Delta\omega_0)$ to the original intensity $I_0(\Delta\omega_0)$ (that is the intensity $I_0(\Delta\omega_0)$ which would be if there were no coupling of the a and a' terms):

$$I(\Delta\omega_0)/I_0(\Delta\omega_0) = |f'_-(R_0)|\{2/[\gamma|f''(R_0)|]\}^{1/2}. \tag{2.4}$$

Here we come to the following *central point*. From equation (2.4) it is clear that if $|f_-'(R_0)|$ was relatively large and/or $\gamma|f''(R_0)|$ was relatively small, we would have $I(\Delta\omega_0)/I_0(\Delta\omega_0) > 1$, thus indicating the formation of a peak (satellite). However, if $|f_-'(R_0)|$ is relatively small and/or $\gamma|f''(R_0)|$ is relatively large, we have $I(\Delta\omega_0)/I_0(\Delta\omega_0) < 1$, thus indicating the formation of a dip, rather than a satellite. This result disproved the existing paradigm in accordance to which any extremum in the transition energy/frequency could manifest only as a satellite.

Specifically, in paper [25] it was shown that if the extremum in the transition energy/frequency is 'sharp' (i.e. $|f''(R_0)|$ is relatively large), then this would manifest as a local dip in the spectral line profile. The absolute value of the second derivative $f''(R_0)$ can be estimated as

$$f''(R_0) \approx \left[f'_+(R_0) - f'_-(R_0)\right]/\delta R, \tag{2.5}$$

where δR is the interval where there occurs the conversion of the original radiator's term a into a'. The interval δR can be found from the following considerations.

Due to the dynamical Stark broadening and the radiative broadening (the latter resulting in a 'natural' width), the radiator has a finite lifetime $1/\gamma$. Consequently, the transition energy of the radiator actually occupies a band of the width γ (see figure 2.1). Therefore, it is easy to find that

$$\delta R = \gamma/|f'_+(R_0) - f'_+(R_0)|. \tag{2.6}$$

Thus the ratio $I(\Delta\omega_0)/I_0(\Delta\omega_0)$ can be represented in the form:

$$I(\Delta\omega_0)/I_0(\Delta\omega_0) \approx 2^{1/2}|f'_-(R_0)/\left[f'_+(R_0) - f'_-(R_0)\right]|. \tag{2.7}$$

Clearly, the right-hand side of equation (2.7) is controlled by the inverse value of the relative change of the derivative of the transition energy at the crossing and does not depend on the dynamical Stark width γ. If the relative change of this derivative is small, there forms a satellite. However, if the relative change of this derivative is large, we find again that the existing paradigm breaks down: there forms a dip (rather than a satellite).

We would like to remind the reader that terms of the quasimolecule ZeZ' may indeed cross: the well-known non-crossing rule [27] is inapplicable because the system possesses an algebraic symmetry higher than the geometrical symmetry [28]. (The corresponding additional conserved quantity is the projection of the generalized Runge–Lenz vector [29] on the internuclear axis.) However, due to the CE some of the crossings transform into avoided crossings [30, 31].

Further, in paper [25] the authors considered a radiative transition in a hydrogen/hydrogen-like atom/ion of the nuclear charge Z at the presence of the nearest perturber which is a fully-stripped ion of the charge $Z' \neq Z$ located at the distance R. The upper (term a) and the lower (term a_0) states involved in the radiative transition have the principal quantum numbers n and n_0, respectively. At some distance R_0, a Z-term of the principal quantum number n (term a) experiences an avoided crossing with a Z'-term of the principal quantum number n' (term a').

In paper [25] it was shown that when an extremum in the transition energy is due to a charge-exchange-caused avoided crossing, which occurs at a relatively large distance

$$R \gg \max(n^2/Z, n'^2/Z'), \qquad (2.8)$$

then practically always it results in a dip in the profile of the corresponding Stark component of the spectral line, rather than in a satellite.

Thus, there are two different mechanisms that explain the formation of X-dips in spectral line profiles: one through the dynamical broadening $\gamma(R) = \gamma_{\mathrm{CE}}(R) + \gamma_{\mathrm{nonCE}}(R)$, another through the behavior of the transition energies $\Delta E(R)$ in the quasimolecule ZeZ'. In paper [32] it was shown these are two independent mechanisms working simultaneously and complementing each other.

The authors of paper [32] also addressed the question of the shape of the X-dip. They showed that the X-dip is a structure consisting of the primary minimum (the dip) surrounded by two adjacent bumps—as it is in the case of the Langmuir-wave-caused dips (L-dips). Practically the most important thing was that, based on the obtained results, the authors of paper [32] provided a *method for determining the rates of CE* from the experimental shape of the X-dips. Below are some details.

We use atomic units and therefore employ the same notation $f(R)$ for both the transition energy and the transition frequency of the radiative transition. We denote as $g(R)$ the area-normalized probability distribution of the internuclear R.

Far from the anticrossing, the lifetime t_{life} of the upper state of the radiator is controlled by the inelastic part of the dynamical broadening by the electron and ion microfields, as well as by the radiative broadening: $t_{\mathrm{life}}(R) = 1/\gamma_{\mathrm{nonCE}}(R)$, where $\gamma_{\mathrm{nonCE}}(R) = \gamma_{\mathrm{Stark}}(R) + \gamma_{\mathrm{rad}}(R)$. Here $\gamma_{\mathrm{Stark}}(R)$ is the frequency of inelastic collisions with electrons and ions leading to virtual transitions from the upper state of the radiator to other states (usually, the dominant contribution to $\gamma_{\mathrm{Stark}}(R)$ is due to electronic collisions), $\gamma_{\mathrm{rad}}(R)$ is the radiative width. The radiative width (which depends on R via the R-dependence of the dipole matrix elements) is expected to be relatively small—so, we would still call the contribution $\gamma_{\mathrm{nonCE}}(R)$ collisional. The quantity $\gamma_{\mathrm{nonCE}}(R)$ varies very slowly away from R_{cr}.

In the vicinity of the anticrossing, there exists an additional channel for the decay of the upper state of the radiator: charge exchange. A sharp decrease of the lifetime due to charge exchange is controlled by the frequency of charge-exchange-causing collisions $\gamma_{\mathrm{CE}}(R) = N_i \langle v\, \sigma_{\mathrm{CE}}(v) \rangle$, where N_i is the density of incident ions, $\langle v\, \sigma_{\mathrm{CE}}(v) \rangle$ is the rate coefficient of CE.

With the allowance for CE, the lifetime of the upper state of the radiator becomes $t_{\mathrm{life}}(R) = 1/\gamma_t(R)$. Here $\gamma_t(R)$ is the total frequency of inelastic collisions equal to $\gamma_{\mathrm{CE}}(R) + \gamma_{\mathrm{nonCE}}(R)$.

Since the lifetime of the upper state of the radiator is at the focus of one of the two mechanisms of the X-dip formation, an appropriate starting formula for the lineshape should be chosen such as to contain explicitly the lifetime of the upper state of the radiator $t_{\text{life}}(R)$ or the total frequency of inelastic collisions $\gamma_t(R)$. It is well-known that the term $\exp[-\gamma_t(R)\tau]$ in the correlation function corresponds to a Lorentzian of the FWHM equal to $\gamma_t(R)$. Therefore, for the area-normalized profile I $(\Delta\omega)$ of the Stark component versus the detuning $\Delta\omega$ from the unperturbed frequency ω_0 we use the following expression

$$I(\Delta\omega) = \int_0^\infty dR \; G(R)L[\Delta\omega - f(R)], \qquad G(R) = g(R)J(R)/J(\infty), \qquad (2.9)$$

where $J(R)$ is a relative intensity of the Stark component and $L(x)$ is the Lorentzian:

$$L(x) = [\gamma_t/(2\pi)]/[(\gamma_t/2)^2 + x^2]. \qquad (2.10)$$

We consider a vicinity of some particular distance R_0 corresponding to a small part of the component profile around $\Delta\omega_0 = f(R_0)$.

In the previous part of this chapter, for deriving the intensity drop at the center of the X-dip, in the integral in equation (2.1) we used the Taylor expansion of the transition energy $f(R)$ at the point $R_0 = R_{\text{cr}}$ corresponding to a possible extremum of $f(R)$ at the anticrossing. Therefore, the term (in the Taylor expansion) proportional to the first derivative $f'(R_0)$ was zero. As a result we obtained:

$$I(\Delta\omega_0) = G_0\{2/[\gamma_t|f_0''|]\}^{1/2}, \qquad G_0 \equiv G(R_0), \qquad f_0'' \equiv f''(R_0). \qquad (2.11)$$

Now we rewrite this result by introducing instead of $\Delta\omega_0$ its scaled (dimensionless) counterpart Ω_0:

$$\Omega_0 \equiv \Omega(R_0) = 2\Delta\omega_0/\gamma_t = 2f(R_0)/\gamma_t. \qquad (2.12)$$

Then the corresponding scaled (dimensionless) intensity $I_{s0}(\Omega_0)$ at the center of the X-dip can be represented as

$$I_{s0}(\Omega_0) \equiv \gamma_t I(\Delta\omega_0)/2 = G_0\{\gamma_t/[2|f_0''|]\}^{1/2}. \qquad (2.13)$$

In distinction to the above, now in the integral in equation (2.1) we expand $f(R)$ at some point R_0 located in the vicinity of the extremum of $f(R)$, but not necessarily coinciding with the point of the extremum. Therefore, the term proportional to the first derivative $f'(R_0)$ will be now different from zero.

We introduce a scaled (dimensionless) counterpart b of this first derivative:

$$b \equiv 2f_0'/(\gamma_t|f_0''|)^{1/2}, \qquad f_0' \equiv f'(R_0). \qquad (2.14)$$

After some elementary transformation, we now obtain the following formula for the scaled intensity $I_s(\Omega_0)$,

$$I_s(\Omega_0) = I_{s0}(\Omega_0)j_{00}(b), \qquad (2.15)$$

where $j_{00}(b)$ is a universal function defined as

$$j_{00}(b) \equiv (2^{1/2}/\pi) \int_{-\infty}^{\infty} dy(1 + 2y^2 + y^4)/[1 + (2 + b^2)y^2$$
$$+ 2by^3 + 2(1 + b^2)y^4 + 4by^5 + (2 + b^2)y^6 + 2by^7 + y^8]. \tag{2.16}$$

The above integral can be calculated analytically. However, the resulting expression is bulky and we do not present it here.

In the previous part of this chapter, the total frequency of inelastic collisions γ_t was considered as a constant (calculated at the center of the dip). Let us relax this assumption.

The two terms within $\gamma_t(R) = \gamma_{CE}(R) + \gamma_{nonCE}(R)$ significantly differ by their dependence on the internuclear distance R: $\gamma_{CE}(R)$ rapidly decreases away from the anticrossing R_{cr}, while $\gamma_{nonCE}(R)$ varies very slowly away from R_{cr}. We introduce the following notations:

$$\gamma \equiv \gamma_{nonCE}(R_{cr}), \quad \gamma_{CEo} \equiv \gamma_{CE}(R_0), \quad a \equiv \gamma_{CEo}/\gamma. \tag{2.17}$$

First, we set $R_0 = R_{cr}$, so that $b = 2f_0'/(\gamma|f_0''|)^{1/2} = 0$ (because $f_0' = 0$ at $R_0 = R_{cr}$), and we can study the second set of questions independently of the first one. By approximating the shape of $\gamma_{CE}(R)$ in the vicinity of R_{cr} as a Lorentzian, after some elementary transformation, we obtain the following formula for the scaled intensity $I_s(\Omega_0)$,

$$I_s(\Omega_0) = I_{s0}(\Omega_0)j_0(a), \tag{2.18}$$

where $j_0(a)$ is a universal function defined as

$$j_0(a) \equiv (2^{1/2}/\pi) \int_{-\infty}^{\infty} dy[1 + a + (2 + a)y^2 + y^4]/[(1 + a)^2$$
$$+ (2 + 2a)y^2 + 2y^4 + 2y^6 + y^8]. \tag{2.19}$$

The above integral can also be calculated analytically, like the integral in equation (2.16). However, the resulting expression is also bulky and we do not present it here.

Now we consider a general case, where both $a = \gamma_{CEo}/\gamma \neq 0$ and $b = 2f_0'/(\gamma|f_0''|)^{1/2} \neq 0$ (we recall that $f_0' \neq 0$ means that $R_0 \neq R_{cr}$). It turns out that in the general case, the scaled intensity $I_s(\Omega_0)$ can be obtained in the form

$$I_s(\Omega_0) = I_{s0}(\Omega_0)j(a, b). \tag{2.20}$$

Here $j(a, b)$ is the following universal function of two variables

$$j(a, b) \equiv (2^{1/2}/\pi) \int_{-\infty}^{\infty} dy\, f(y, a, b), \tag{2.21}$$

where

$$f(y, a, b) \equiv [1 + a + (2 + a)y^2 + y^4]/[(1 + a)^2 + (2 + 2a + b^2)y^2 \\ + 2by^3 + 2(1 + b^2)y^4 + 4by^5 + (2 + b^2)y^6 + 2by^7 + y^8]. \tag{2.22}$$

The universal function $j(a, b)$ reduces to $j_{00}(b)$ for $a = 0$ and to $j_0(a)$ for $b = 0$. It is the even function of b: $j(a, -b) = j(a, b)$.

For studying a *relative* variation of $j(a, b)$, it is convenient to introduce another function:

$$h(a, b) \equiv j(a, b)/j(a, 0) - 1. \tag{2.23}$$

For any given a, it shows a relative change of the intensity away from the center of the X-dip.

For any given $a = \gamma_{\mathrm{CEo}}/\gamma$, we call the maximal value of $h(a, b)$ as the bump-to-dip contrast (BDC). Figure 2.2 shows the BDC versus a.

Thus, using the dependence of the BDC versus $a = \gamma_{\mathrm{CEo}}/\gamma$, *we can determine the ratio $\gamma_{CE}(R_0)/\gamma$ from the experimentally measured BDC*. We note that

$$\gamma_{\mathrm{CEo}} = N_i \langle v \, \sigma_{\mathrm{CE}}(v) \rangle, \tag{2.24}$$

where N_i is the density of incident ions, $\langle v \, \sigma_{\mathrm{CE}}(v) \rangle$ is the rate coefficient of CE. Consequently, after determining the experimental value $a_{\mathrm{exp}} = \gamma_{\mathrm{CEo}}/\gamma$ from the experimental BDC, *we can then deduce the rate coefficient of CE as follows*:

$$\langle v \, \sigma_{\mathrm{CE}}(v) \rangle = \gamma a_{\mathrm{exp}}/N_i. \tag{2.25}$$

The quantity $\gamma \equiv \gamma_{\mathrm{nonCE}}(R_{\mathrm{cr}})$ in equation (2.25), representing the frequency of inelastic collisions with electrons and ions leading to virtual transitions from the upper state of the radiator to other states, can be calculated for given plasma parameters N_e, N_i, T_e, and T_i by using one of few contemporary theories (presented, e.g. in book [34], see also appendix A of the present book). The experimental determination of the rates of CE for multicharged ions from using X-dips is an important reference data virtually inaccessible by other experimental methods.

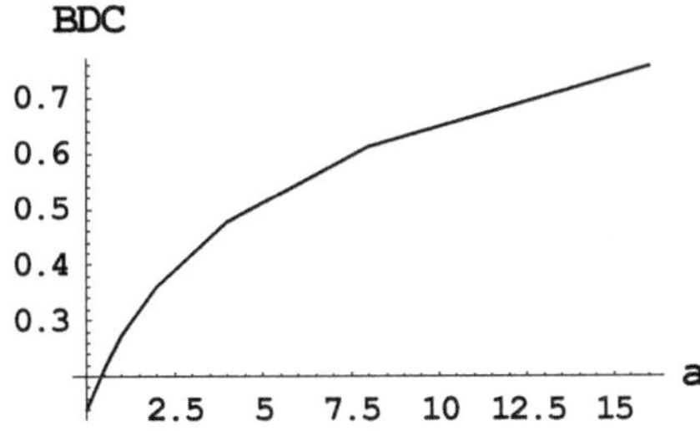

Figure 2.2. Bump-to-dip contrast (BDC) versus the dimensionless parameter a (defined by equation (2.17)). The BDC is defined as the maximal value of the function $h(a, b)$ for any given value of a. (Reproduced with permission from [33]. Copyright 2015 World Scientific.)

2.2 Discovery of the charge-exchange-caused dips in experimental spectral line profiles from laser-produced plasmas

The experiment [35], where the X-dips in profiles of spectral lines emitted by laser-produced plasmas were first discovered, has been performed at the nanosecond laser facility at LULI, France. A laser beam (4ω) of the intensity 4×10^{14} W cm^{-2} in a pulse of 500 ps was focused onto a target. The targets were structured: powdered aluminium carbide (Al_4C_3) strips of the thickness 20 μm were inserted in a carbon substrate. For each target, the Al_4C_3 strip was placed according to the 80 μm diameter of the focal spot. A high-resolution vertical-geometry Johann spectrometer ($R = 8000$) was used for observing the Lyγ line of Al XIII ($Z = 13$) perturbed by fully stripped carbon CVI ($Z'' = 6$).

The experimental profiles exhibited dips are located at 6 mÅ and 9 mÅ from the center of the line. These positions were close to the predicted positions of the X-dips (6.7 mÅ and 9.6 Å, respectively) calculated analytically in paper [36]. The positions of the experimental dips did not vary significantly in the electron density $N_e = 10^{20}$–10^{22} cm^{-3}, thus additionally reinforcing their identification as the X-dips: according to the theory, the positions of the X-dips have only a very weak dependence on the electron density.

For smaller electron densities the spectral line profile was too narrow to allow the visibility of the X-dip structures. At higher densities the dips were smoothed out by the Stark broadening. All these features are illustrated in figure 2.3.

The next experiment [32], where the X-dips were observed in spectral line profiles emitted by laser-produced plasmas, was performed at the same facility as experiment [35]. The setup and the geometry of the experiment are shown in figure 2.4. A single laser beam delivering 5–10 J of frequency-quadrupled radiation (0.263 μm) in a pulse length of 0.5 ns was focused onto a 100 μm diameter spot with an intensity of 2×10^{14} W cm^{-2}. The radiation was incident perpendicularly to the target. The target was structured: the (20–60) μm-thick powdered aluminum carbide Al_4C_3 strip sandwiched between plastic substrate of thickness 20–40 μm. Because the Al_4C_3 strip was placed through the center of the focal spot, the plasma gradients in the transverse directions were suppressed. The aluminum ions were kept in a constrained flow perpendicular to the direction of the spectra observation. This led to the most favorable conditions for observing the X-dips.

The emitted radiation was analyzed by two x-ray spectrometers. A classic Johann spectrometer, having a cylindrically bent crystal of PET, monitored the plasma emission within the range of 5.3–7.4 Å: it monitored the entire Lyman series of Al XIII and spectral series of Al XII (excluding the He-alpha line) were covered. The spatial resolution was provided by a 20 μm-wide slit achieving the transverse magnification of about 100. The high-dispersion spectroscopic data, required for the reliable identification of the X-dips in the line profiles, were collected by a vertical-geometry Johann spectrometer (VJS). The instrument was fitted with a crystal of quartz (100) cylindrically bent to the radius 77.2 mm. The ray-tracing calculations yielded the spectral resolution of 4200. The spectra were recorded on

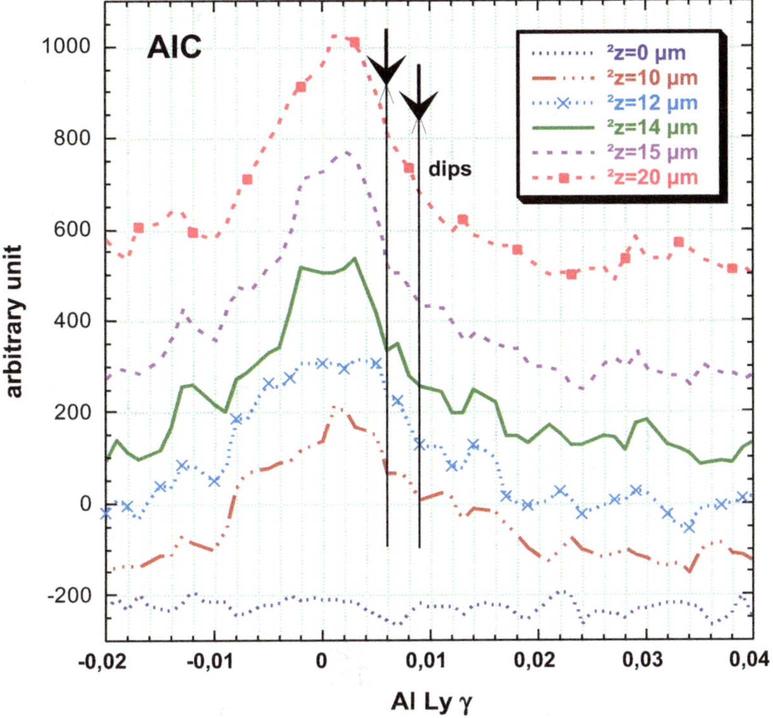

Figure 2.3. Experimental profiles of the Ly-gamma line of Al XIII emitted from an aluminum carbide plasma [35]. The horizontal axis is $\Delta\lambda$ in Angstrom. The evolution of the spectra is given as the space integration Δx along the laser target axis increases. Progressively the average density of the emitting plasma is decreasing. When Δx reaches 20 µm, the average density gets rapidly lower, the densest part of the plasma being hidden by the slit (15 µm). The profiles exhibit two pronounced dips in the red wing for the densities 10^{20}–10^{22} cm^{-3}.

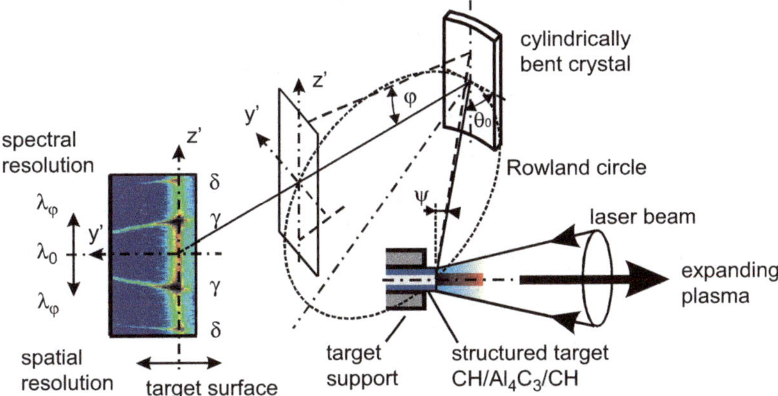

Figure 2.4. Experimental setup showing the structured target (CH/Al$_4$C$_3$/CH or Mg/Al/Mg), the vertical-geometry Johann spectrometer (VJS), and a sample experimental record consisting of two identical (except a noise) sets of spatially resolved spectra symmetrically located with respect to the central wavelength λ_0 [32].

x-ray film Kodak Industrex CX. The spatial resolution was 8 µm—determined by the microdensitometer slit.

The simultaneous production of a pair of symmetric spectra using the VJS provided a reference point λ_0 for the computational reconstruction of the raw spectroscopic data and, at the same time, considerably enhanced the reliability of identifying the X-dips.

The information about the spatial distribution of the plasma characteristics relevant for time-integrated x-ray emission of hydrogenic ions was obtained from the modified one-dimensional (1D) hydrodynamic code MEDUSA [37]. The results—in terms of the electron densities—were compared with the corresponding results deduced from the analysis of the shape of the experimental Al Ly-gamma profiles. For this purpose, for higher electron densities the authors used the code IDEFIX [38, 39]. For lower electron densities they utilized the code PIM PAM POUM [40]. These two sets of the deduced electron densities were in agreement within approximately 20%. The 20% difference was due to approximations used in modeling and to experimental errors of the spectral line measurements.

Figure 2.5 presents four experimental profiles of the Al XIII Ly-gamma line obtained with the help of the VJS. For these profiles, the analysis of the broadening of the Ly-gamma line yielded the electron densities 3.0×10^{22} cm^{-3}, 2.2×10^{22} cm^{-3}, 1.1×10^{22} cm^{-3}, and 0.5×10^{22} cm^{-3} for the profiles 1, 2, 3, and 4, respectively. The observed line profiles are more structured. This is because the X-dips are structures where the primary minimum is surrounded by two bumps.

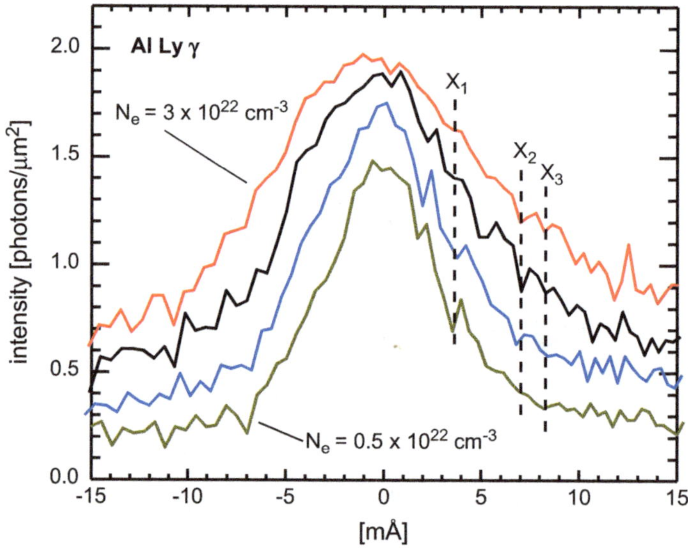

Figure 2.5. Experimental profiles of the Ly-gamma line of Al XIII emitted from Al$_4$C$_3$ target sandwiched between CH substrates [32]. Profiles from the top to the bottom correspond to the electron densities 3.0×10^{22} cm^{-3}, 2.2×10^{22} cm^{-3}, 1.1×10^{22} cm^{-3}, and 0.5×10^{22} cm^{-3}, respectively. Dashed vertical lines mark calculated, density-independent positions of the X-dips.

The positions of the X-dips (the positions of their central minima) do not depend significantly on the electron density for our experimental density range. We marked (by vertical dashed lines) the positions x_1, x_2, and x_3 ($\Delta\lambda = 3.2$ mÅ, 6.7 mÅ, and 9.6 mÅ) calculated analytically by the theory from paper [36]—the *same positions for all four experimental profiles*. All predicted X-dips, albeit tiny, are well visible and can be found in each experimental profile.

The authors of paper [32] noted that bump positions slightly vary with the density. Indeed, according to the theoretical part of paper [32] (presented also in the previous section), bump positions vary with the parameter $a \equiv \gamma_{\text{CEo}}/\gamma$. While the quantity γ_{CEo} scales linearly with the density, the scaling of the quantity $\gamma \equiv \gamma_{\text{nonCE}}(R_{\text{cr}}) = \gamma_{\text{Stark}}(R_{\text{cr}}) + \gamma_{\text{rad}}(R_{\text{cr}})$ with the density is slightly nonlinear. The latter is due to three reasons. First, the electron (dominant) contribution to $\gamma_{\text{Stark}}(R_{\text{cr}})$ has a slightly-nonlinear density scaling (see, e.g. book [34]). Second, the density scaling of the ionic contribution to $\gamma_{\text{Stark}}(R_{\text{cr}})$ has an even stronger non-linearity than the scaling of the electronic contribution [34]. Third, the relatively small contribution from the radiative width $\gamma_{\text{rad}}(R_{\text{cr}})$ is practically independent of the density—therefore, even if $\gamma_{\text{Stark}}(R_{\text{cr}})$ had scaled linearly with the density, the ratio $\gamma_{\text{CEo}}/[\gamma_{\text{Stark}}(R_{\text{cr}}) + \gamma_{\text{rad}}(R_{\text{cr}})]$ would still slightly depend on the density.

The slight density dependence of the bump positions did not affect the experimental determination of the rate coefficient of CE $\langle v\, \sigma_{\text{CE}}(v)\rangle = \gamma a_{\text{exp}}/N_{\text{i}}$. The reason is that the slightly nonlinear density scalings of γ and of a_{exp} exactly compensate each other: so, the product γa_{exp} scales linearly with the density. Indeed, $\gamma a_{\text{exp}} = \gamma_{\text{CEo}}$ and the quantity γ_{CEo} has a linear density scaling.

The authors of paper [32] specified three reasons why the observed dips correspond to real phenomena. First, the features were reproducibly observed on both symmetric spectral lineouts (simultaneously produced by the VJS) at the same positions. Second, the modulation of the intensity by the most pronounced bump–dip–bump structures is about 10%, while the standard deviation of the intensity measurement at the locations of these dips is between 3% and 4%. Third, such dips have not been observed when the authors changed the target by using Al strip instead of Al_4C_3 strip: in the experiments with the Al target they cannot occur—in accordance with the theory.

Figure 2.6 shows a magnified part of the experimental dip x_1 from the bottom profile in figure 2.5. This spectrum corresponds to the emission from a plasma region of the electron density 0.5×10^{22} cm^{-3} and of the temperature (500–600) eV. This experimental X-dip was used by the authors of paper [32] for the determination of the rate coefficient of CE.

The experimental dip x_1 shown in detail in figure 2.6 is well isolated from the dips x_2 and x_3. Therefore, it is best suited for the experimental determination of the bump-to-dip contrast (BDC).

The experimental determination of the BDC from the part of the experimental profile in the vicinity of the dip x_1 consists of the following steps: (1) obtaining the 'unperturbed' (no X-dip) profile by smoothing the experimental profile (using the standard technique) until the X-dip disappears; (2) correcting the relative differences of the dip and bumps intensities on the inclination of the unperturbed experimental

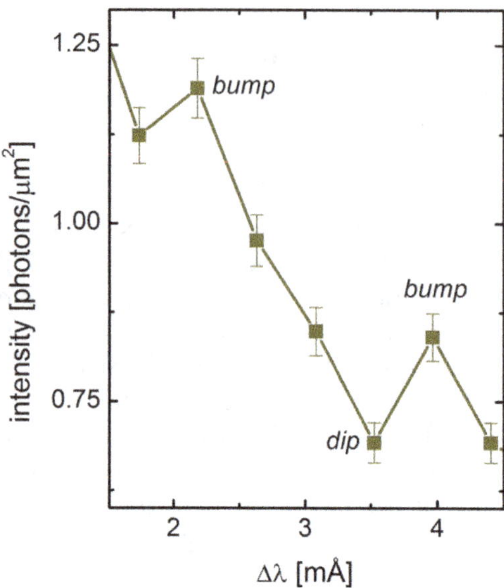

Figure 2.6. Magnified part of the experimental dip x_1 from the bottom profile in figure 2.6 [32].

profile; (3) calculating the BDC from the corrected relative differences of the dip and bumps intensities. By doing so, the authors of paper [32] obtained the experimental value of the BDC equal to 0.34 ± 0.02 (as a result of the averaging over the two bumps). From this value of the BDC, using the theoretical dependence of the BDC on the parameter $a = \gamma_{CE_0}/\gamma$ (see figure 2.2), they deduced the experimental value of $a_{exp} = 1.8 \pm 0.2$. Finally, by substituting in equation (2.25) the experimental values of a_{exp} and N_i, as well as the calculated value of γ [34], the authors of paper [32] found the rate coefficient of CE between the hydrogenic aluminum in the state of $n = 4$ and a fully-stripped carbon to be

$$\langle v\, \sigma_{CE}(v)\rangle = (5.2 \pm 1.1) \times 10^{-6}\ \mathrm{cm^3\, s^{-1}}. \tag{2.26}$$

Thus, the authors of paper [32], by applying their theory to their experimental data, determined for the first time the rate coefficient of CE between the hydrogenic aluminum and a fully-stripped carbon in the laser-produced plasma. This constituted the next step toward the employment of the X-dip phenomenon for producing not-yet-available fundamental data on CE between multicharged ions, virtually inaccessible by other experimental methods.

The next experimental study of the X-dips was performed in plasma–wall interaction experiments at the PALS facility—see paper [41] and review [42]. A plasma jet of aluminum ions was produced by the nanosecond iodine laser. The laser intensity was 3×10^{14} W cm^{-2}. The laser radiation was incident on a foil and interacted with a massive carbon target. The study employed the high spectral and spatial resolution vertical-geometry Johann spectrometer. It was used for observing profiles of the Al XIII Ly-gamma line.

Figure 2.7 presents the experimental profiles exhibiting two X-dips in the red wing. These X-dips are clearly visible in the experimental profiles corresponding to sufficiently high electron densities. (According to simulations, the electron density could reach 5×10^{22} cm^{-3}.)

The latest experimental study of X-dips, performed at Kansai Photon Science Institute of Japan Atomic Energy Agency, was quite unique [43]. To begin with, this was the first study of L-dips and X-dips in spectral lines from *femtosecond* laser-driven *cluster-based* plasma. Second, in distinction to the previous experimental studies of the X-dips and L-dips, the authors employed a spectral line of an ion different from Al XIII, namely the Ly-epsilon line of O VIII. Third, but perhaps most importantly, the observed L-dips turned out to be caused by Langmuir waves at the frequency $\omega_p(N_e) = \omega_{\mathrm{las}}/2$, hence corresponding to the electron density equal to one quarter of the critical density. Thus, these Langmuir waves resulted from the well-known parametric instability, namely the two-plasmon decay instability, so that the experiment [43] constituted the first observation of the signature of this instability in spectral line profiles. Fourth, the authors of paper [43] also observed an X-dip and used it for the experimental determination of the rate of CE between the hydrogenic oxygen and fully-stripped helium (helium was one of the plasma components).

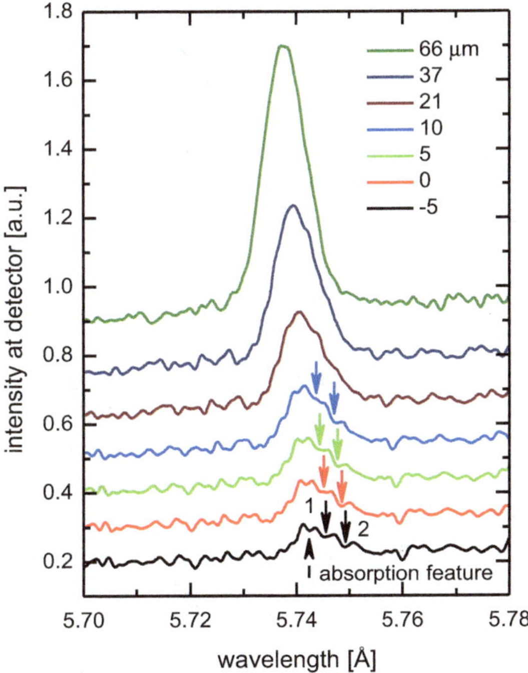

Figure 2.7. Observation of the X-dips in the red wing of the Ly-gamma line of Al XIII in the plasma–wall interaction experiments at the PALS laser facility. A plasma jet of aluminum ions interacted with a massive carbon target. The spectra correspond to different distances from the target surface (and thus to different electron densities): from –5 µm (the bottom curve) to +66 µm (the top curve) [41, 42].

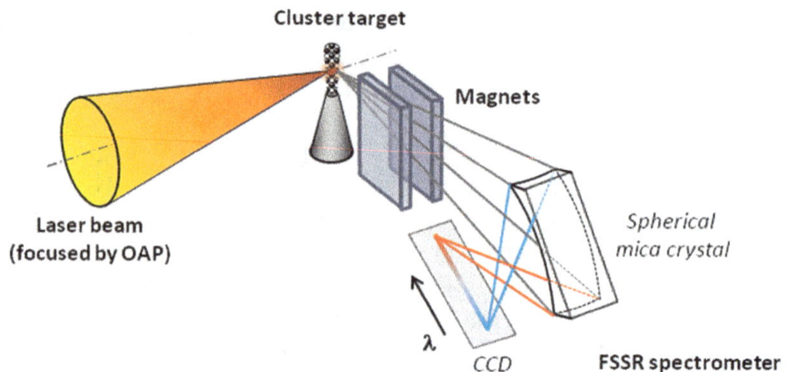

Figure 2.8. Scheme of the experiments [43].

Here are some additional details from the theory of X-dips (see, e.g. book [34], section 8.2), required for understanding the results from paper [43]. An X-dip originates from CE occurring at an anticrossing of two terms of ZeZ' system. According to a selection rule, the total number ν of anticrossings is $\nu = \min(2n - 1, 2n' - 1)$. Here n and n' are the principal quantum numbers labeling the two terms at large internuclear distances R. For X-dips to be observed, the ratio n/n' should be relatively close—but not too close—to the ratio Z/Z'.

Now we can explain a unique opportunity found by the authors of paper [43]. For the quasimolecule ZeZ', where $Z = 8$ (O^{+8}) and $Z' = 2$ (He^{+2}), the optimal choice of the principal quantum numbers is $n = 6, n' = 2$ (thus, the choice of the spectral line of O VIII is Ly-epsilon). This means that the corresponding terms have three anticrossings. Generally, each anticrossing corresponds to a possible X-dip at a different location in the line profile. However, for this particular case as an exception, all three anticrossings correspond to three possible X-dips practically at the same location in the profile of the line Ly_ε of O VIII—merging into one X-superdip located at (30 ± 3.5) mA in the red wing.

This explains the choice of the plasma composition and of the particular spectral line of O VIII in experiment [43]. The experimental results far exceeded the expectations: the authors of paper [43] observed not only the X-superdip, but also L-dips. The latter allowed an accurate experimental determination of the electron density (as well as of the amplitude of the electric field of the Langmuir wave) leading to the conclusion that they observed the spectroscopic signature of the two-plasmon decay instability.

In the experiments presented in paper [43], two Ti:sapphire laser facilities (wavelength approximately 800 nm) were used. In the first experiment the JLITE-X laser generated 40 fs pulses of the energy of 160 mJ with the contrast of 10^5. The laser beam was focused about 1.5 mm above the nozzle orifice by an off-axis parabola with the spot size of around 50 μm, which yields the laser intensity of 4×10^{17} W cm^{-2} in vacuum (figure 2.8). In the second experiment the J-KAREN laser provided the laser pulses with a high contrast of 10^8–10^{10}, achieved using an additional saturable absorber and an additional Pokkels cell switch. The pulse duration was 40 fs at the pulse energy of about 800 mJ and the intensity of laser radiation in the focal spot with a diameter of 30 μm reached 3×10^{18} W cm^{-2} in vacuum.

2D hydrodynamic calculations have been performed to design supersonic nozzle having a three-staged conical structure for the purpose of producing a sufficient amount of submicron-sized clusters at room temperature [44]. A pulsed solenoid valve connected to such a nozzle has an entrance orifice diameter of 0.5 mm and an output orifice diameter of 2 mm. The opening time of the solenoid pulsed valve connected to the nozzle was set at 1 ms to ensure a stable gas flow. The clusters were created when a gas of a high initial pressure was expanded into vacuum through a nozzle. It allowed producing CO_2 clusters with a diameter of about 0.5 ± 0.1 μm for pure CO_2 (gas pressure before expansion was 20 bar) and 0.26 ± 0.04 μm for the mixed gas of 90% He + 10% CO_2 (gas pressure before expansion was 60 bar).

The spatially resolved x-ray spectra have been recorded by employing a focusing spectrometer with spatial resolution-1D (FSSR-1D) [45, 46]. In such a scheme, experimental data are acquired by placing the detector on the Rowland circle of a spherically bent crystal. The spectral resolution of the spectrometer does not depend on the size of the plasma source. This spectrometer was equipped with a spherically bent mica crystal ($R = 150$ mm) and a vacuum compatible x-ray charge coupled device (CCD) camera (DX440-BN, Andor, ME) with a pixel size of 13.5 mm. The spectral resolution was up to $\lambda/\Delta\lambda \sim 4000$. A magnet was placed between the plasma source and the crystal in order to stop high-energy particles. A typical spectrogram obtained with our focusing spectrometer is shown in figure 2.9.

Figure 2.9 shows four experimental profiles of the O VIII Ly-epsilon line obtained during four different laser shots. In two of these cases (labeled #1 and 2) the laser irradiated a mixture of CO_2 and He, while in the two other cases (labeled #3 and 4) the laser irradiated just CO_2. In all four profiles, the slightly shifted position of the center of the line is marked by a dash-dotted vertical line.

Profile #1 corresponds to the highest laser intensity: 3×10^{18} W cm^{-2}. Two solid vertical lines correspond to the positions of two L-dips: one dip in the blue wing at -20 mA from the slightly shifted center of the line, another L-dip in the red wing at 37 mA from the slightly shifted position of the center line, as shown in figure 2.9 inset. The center of gravity of the two L-dips is shifted to the red by 9 mA, which according to [47] is because at high electron densities, the multipole interactions higher than the dipole interaction are significant.

The superposition of a bump–dip–bump structure in the blue wing with a significantly inclined spectral profile created a secondary minimum at about 14.63 A of no physical significance. As for the L-dip in the red wing, its near bump is clearly visible, but the far bump is only faintly outlined because it practically merged with the noise. Here and below the 'near' (or 'far') bump means the bump closer to (or further from) the line center with respect to the central minimum of the dip.

The two L-dips in profile #1 are separated from each other by $4\lambda_p$, where $\lambda_p = \omega_p \lambda_0^2/(2\pi c)$. They are one-quantum resonance dips ($s = 1$ in $\omega_F = s\omega_p(N_e)$) in the profiles of the two most intense lateral components of the Ly$_\varepsilon$ line, originating from the Stark sublevels (311) and (131), the sublevels being labeled by the parabolic

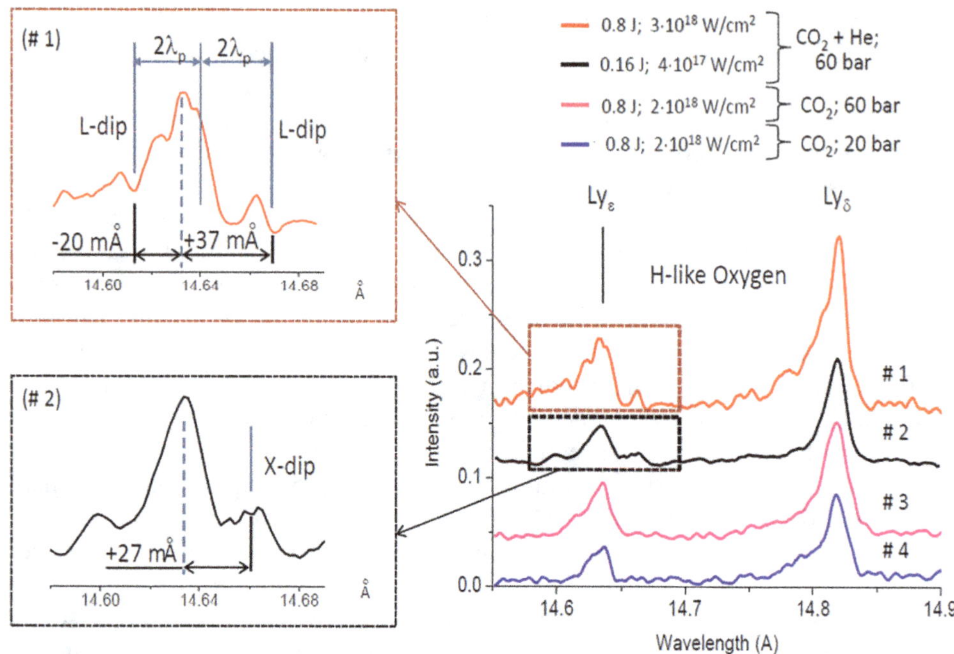

Figure 2.9. Typical spectra of the O VIII Ly$_\delta$ and Ly$_\varepsilon$ lines obtained in femtosecond laser-driven cluster-based experiments [43]. Enlarged spectra of the O VIII Ly$_\varepsilon$ line in the inset show the positions of the L- and X-dips in more detail.

quantum numbers. The electron density deduced from the separation of the two L-dips was $N_e = 5.0 \times 10^{20}$ cm^{-3}. This value of N_e is quite reasonable for our experimental conditions. Indeed, the density of atoms and molecules in the gas jet is about 3×10^{19} cm^{-3}, and at a high laser intensity of 3×10^{18} W cm^{-2}, the optical field ionization ionized the helium atoms completely, while carbon and oxygen up to Li-like ions. The collisional mechanism additionally ionized such ions up to fully-stripped ones, which follows from observing highly-excited Lyman lines of O VIII. Thus, multiplying 3×10^{19} cm^{-3} by 22 (the number of electrons in CO_2 molecules), the authors of paper [43] expected that the maximal electron density in their experiments could be up to 6.6×10^{20} cm^{-3}.

At such a high electron density, at the absence of L-dips, the profile of the Ly-epsilon line might have looked like a triplet. In addition to the primary maximum (representing mostly the profile of the central, 'unshifted' Stark component), the lateral maxima (one on each side) would represent the combined profile of the lateral Stark components. However, at such a high electron density, the multipole interactions higher than the dipole interaction become very important, enhancing the red lateral maximum and suppressing the blue one [48, 49]. This explains a small 'shoulder' at about 14.64 Å. The electron density estimated from the position of this shoulder is in agreement with the electron density deduced more accurately from the L-dips positions.

The value of the plasma electron frequency ω_p deduced from the separation of the experimental L-dips was $\omega_p = 1.26 \times 10^{15}$ s^{-1}, while the laser frequency was $\omega_{\mathrm{laser}} = 2.4 \times 10^{15}$ s^{-1}. It is seen that $\omega_{\mathrm{laser}} = 2\omega_p$ (within the accuracy of 5%). This means that the authors of paper [43] observed in the spectral line profile a signature of the two-plasmon decay instability. This parametric instability occurring at the quarter-critical density is one of the major processes characterizing laser–plasma interaction (see, e.g. [50–53]). It results in the decay of one quantum of the laser radiation into two Langmuir quanta (plasmons).

So, in shot #1 the interaction of laser radiation of the very high laser intensity 3×10^{18} W cm^{-2} with CO_2–He mixture caused strong Langmuir waves in the corresponding plasmas—strong enough to produce pronounced L-dips. In distinction, in shot #2 the laser intensity was by an order of magnitude smaller (4×10^{17} W cm^{-2}). Therefore, in the corresponding plasma, Langmuir waves (if any) were significantly weaker and did not manifest as identifiable L-dips in the profile.

From the halfwidth of the L-dips in profile #1, the authors of paper [43] found the amplitude of the electric field of the Langmuir wave. It was $E_0 = 40$ MV cm^{-1}.

The theory predicted also a possible X-superdip at (30 ± 3.5) mA in the red wing of the O VIII Ly-epsilon line. This X-dip is clearly visible in profile #2: it is marked by the solid vertical line in the inset in figure 2.9. The central minimum of the experimental X-dip is at 27 mA (in good agreement with the theory) and is surrounded by two bumps, as predicted by the theory. In the profile from shots #1, the possible X-dip at 27 mA was practically 'swallowed' by the near bump of the L-dip. More accurately, in the profile from shots #1, the bump between the central minimum of the L-dip at 37 mA and the central minimum of the X-dip at 27 mA is a superposition of the near bump of the L-dip and of the far bump of the X-dip, which is why this bump is so intense.

As for shots #3 and #4 without the admixture of He, there was no X-dip at 27 mA. This was expected for the plasma without He and thus further confirmed the identification of the X-dip in CO_2–He mixture. Whether or not there were Langmuir waves in the plasma from shots #3 and #4 was hard to determine reliably because the modulations of the profile at the possible locations of the L-dips were the level of the noise. The authors of paper [43] noted that the shot #1, where the L-dips were reliably identified, had the laser intensity by 50% greater than shots # 3 and #4.

From the experimental bump-to-dip-contrast (BDC) of the X-dip in profile #2, the authors of paper [43] obtained the rate of CE between the ion O^{+8} in the state of $N = 6$ and the ion He^{+2} in the state of $n'' = 2$:

$$\langle v \, \sigma_{\mathrm{CE}}(v) \rangle = (1.5 \pm 0.3) \times 10^{-6} \; \mathrm{cm^3 \, s^{-1}}. \tag{2.27}$$

This is new fundamental data virtually inaccessible by other experimental methods. The error bar of this data was due to the combined result of both the relative inaccuracy in measuring a_{exp} and the uncertainty of the temperature ($T_e \sim 110$ eV was estimated independently using the intensity ratio of the O VIII Ly$_\alpha$ and O VII He$_\beta$ lines).

2.3 Theory of the charge-exchange-caused dips in profiles of He-like spectral lines

In paper [54], which we follow below, it was shown that X-dips are possible also on the profiles of He-like spectral lines. In the previous studies the X-dip phenomenon was considered to be possible only in spectral lines of hydrogen atoms and hydrogen-like ions: due to the existence of *exact* algebraic symmetries relevant only to hydrogenic systems and to the corresponding two-Coulomb-center systems having one electron. Exact algebraic (i.e. higher than geometrical) symmetries of these quantum systems lead to exact additional conserved quantities, known as Runge–Lenz vector and its generalizations (see,. e.g. the latest paper [29] on this subject).

The concept of symmetry is very general: in fact, any approximate analytical theory can be considered as a simplified version of a more complicated problem, the simplification being achieved by using some *approximate* symmetry [55]. Actually, any regularity in the energy spectrum of a quantum system reflects certain symmetry properties [55].

While helium-like ions and the corresponding two-Coulomb-center systems having two electrons do not possess *exact* additional conserved quantities, they possess *approximate* additional conserved quantities [56]. This should be sufficient for the X-dips to occur also in these systems.

In more detail, the higher than geometrical symmetry of two-Coulomb-center systems having one electron is manifested by an additional conserved quantity (integral of motion)—in addition to the energy E and the angular momentum projection M. This additional conserved quantity is the projection (on the internuclear axis) of the super-generalized Runge–Lenz vector [29]

$$\mathbf{A} = \mathbf{p} \times \mathbf{L} - L^2/R\, \mathbf{e}_z - Z\mathbf{r}/r - Z'(\mathbf{R} - \mathbf{r})/|\mathbf{R} - \mathbf{r}| + Z'\mathbf{e}_z,\ \mathbf{e}_z = \mathbf{R}/R, \qquad (2.28)$$

where $\mathbf{p}$ and $\mathbf{L}$ are the linear and angular momenta vectors, respectively; $\mathbf{r}$ is the radius vector of the electron. In the two-electron case, after substituting in equation (2.28) Z by $Z_{\mathrm{eff}} = Z - 1$ and treating $\mathbf{r}$ as the radius vector of the outer electron, the projection of the vector $\mathbf{A}$ on the internuclear axis is an approximately conserved quantity.

This analytical theory, as well the corresponding analytical theory for the one-electron case presented in section 2.1, is valid for the typical situation where the internuclear distance R_{c}, at which a quasicrossing occurs, satisfies the following conditions. The first condition is

$$R_c \gg \max(n^2/Z_{\mathrm{eff}},\ n'^2/Z'), \qquad (2.29)$$

i.e. for R_c to significantly exceed the characteristic sizes of the two corresponding hydrogenic subsystems. Under this condition, one can use the multipole expansion for the energy terms. The second condition is

$$R_c > \max[(3n^5 Z'/Z_{\mathrm{eff}}^3)^{1/2},\ (3n'^5 Z_{\mathrm{eff}}/Z'^3)^{1/2}]. \qquad (2.30)$$

This condition is necessary to ensure the existence of the energy level n_{eff} of the radiating Z_{eff}-ion and of the energy level n' of the Z'-ion. Namely, relation (2.30) is necessary to ensure that under the electric field of Z'-ion at the distance R_c, the level n of the radiating Z_{eff}-ion does not merge with the level $n + 1$, as well as to ensure that under the electric field of Z_{eff}-ion at the distance R_c, the level n' of the Z'-ion does not merge with the level $n' + 1$. For practically all dicenters, the condition (2.30) is more restrictive than the condition (2.29).

The third condition puts an upper limit on R_c:

$$R_c < [3n^2 Z'/(Z_{\text{eff}} \, \Delta E)]^{1/2}, \qquad (2.31)$$

where ΔE is the size of the unperturbed multiplet of the principal quantum number n. This inequality ensures that for the Z-ion state of the principal quantum number n, the Stark splitting caused by the Z'-ion at the distance R_c significantly exceeds the unperturbed separation of the sublevels of the n-shell. This condition allows using parabolic coordinates.

It can be shown that for the restrictions on R_c to be fulfilled, the ratio Z_{eff}/Z' should slightly exceed (but not be equal to) a small integer. This explains the choice of prospective candidates for observing X-dips in spectral line profiles of He-like ions from laser-produced plasmas presented below.

In paper [54] the authors identified three prospective candidates for observing X-dips in spectral line profiles of He-like ions from laser-produced plasmas, as presented below. Analytical calculations of the positions of the X-dips in spectral lines of He-like ions require, in particular, analytical calculations of the effective principal quantum number n_{eff}, which in its turn requires analytical calculations of the quantum defect of the energy levels of He-like ions. For this purpose in paper [54] the authors used the analytical method developed by Nadezhdin and Oks [57]. This method (just like other methods, to the best of our knowledge) calculates the quantum defect in the spherical quantization. However, under the condition (2.31) the Stark effect is linear, which means also that the relatively strong electric field of Z'-ion at the distance R_c from the radiating Z_{eff}-ion intermixes all the spherical eigenfunctions of the energy level n. Therefore, the effective principal quantum number n_{eff}, used then in the parabolic quantization's formula, should be calculated by averaging the quantum defect over all sublevels of the spherical quantization belonging to the level n.

For each of the three prospective dicentres presented below, the authors of paper [54] compared the energy, obtained from the averaging Nadezhdin–Oks quantum defect, with the corresponding result that can be deduced from the empirical data from paper [58]. It turned out that for these and other dicentres, already for $n = 4$ used in the examples, the relative difference was only $\sim 0.05\%$. Such an excellent agreement justifies the usage of the quantum defect concept for $n = 4$. Of course, for higher values of n, the accuracy would be even better.

For the examples given below, all three conditions (2.29)–(2.31) are satisfied.

1. He-gamma line of Si XIII 5.405 A ($Z = 14$) perturbed by fully-stripped C ($Z' = 6$). For the corresponding experiment the solid target can be made out of silicon carbide (SiC), also known as carborundum.
2. He-gamma line of S XV 4.0885 A ($Z = 16$) perturbed by fully-stripped N ($Z' = 7$). For the corresponding experiment the solid target can be made out of crystals/granules of ammonium sulfate $(NH_4)_2SO_4$.
3. He-gamma line of Mg XI 7.4731 A ($Z = 12$) perturbed by fully-stripped B ($Z' = 5$). For the corresponding experiment the solid target can be made out of crystals of magnesium borate $(MgO)_3B_2O_3$.

In each of the above three examples, there are quasicrossings of some of the Z_{eff}-terms of $n = 4$ with some of the Z'-terms of $n' = 2$. More specifically, there are the following three avoided crossings:

(1) (Z, n)-term of $q = -3$ with the (Z', n')-term of $q' = -1$;
(2) (Z, n)-term of $q = -2$ with the (Z', n')-term of $q' = 0$;
(3) (Z, n)-term of $q = -1$ with the (Z', n')-term of $q' = 1$.

Thus, there could be up to three X-dips observed in the profile of the He-gamma line of the above three radiating ions. Below we will label the three possible X-dips according to the above order of quasicrossings: the first dip results from the first quasicrossing, the second dip from the second quasicrossing, the third dip from the third quasicrossing.

However, it should be noted that it would be unlikely to observe all three X-dips in one experiment because the range of plasma parameters, favourable for observing an X-dip, differs for the above three X-dips. It depends on the relation between the most probable distance between the radiating and perturbing ions (which depends on plasma parameters) and the distance where the corresponding quasicrossing occurs. Therefore, most probably in some experiments only the X-dips #1 and #2 could be observed, while in experiments at another range of plasma parameters only the X-dips #2 and #3 could be observed. This situation is similar to the X-dips actually observed in the Ly-gamma line of Al XIII ($Z = 13$) perturbed by fully-stripped C ($Z' = 6$), as reported in two different ranges of the experimental conditions [41].

For the three chosen examples of the dicentres, the detailed results can be presented as follows. All the positions of the X-dips are in the red part of the corresponding spectral line profile.

For the He-gamma line of Si XIII 5.405 A ($Z = 14$) perturbed byfully-stripped C ($Z' = 6$):

- the first dip is at 9.0 mA, corresponding to the crossing at $R = 8.3$ au;
- the second dip is at 6.3 mA, corresponding to the crossing at $R = 7.7$ au;
- the third dip is at 3.0 mA, corresponding to the crossing at $R = 7.2$ au.

For the He-gamma line of S XV 4.0885 A ($Z = 16$) perturbed by fully-stripped N ($Z' = 7$):
- the first dip is at 5.3 mA, corresponding to the crossing at $R = 8.1$ au;
- the second dip is at 3.7 mA, corresponding to the crossing at $R = 7.7$ au;
- the third dip is at 1.7 mA, corresponding to the crossing at $R = 7.3$ au.

For the He-gamma line of Mg XI 7.4731 A ($Z = 14$) perturbed by fully-stripped B ($Z' = 5$):
- the first dip is at 16.5 mA, corresponding to the crossing at $R = 8.5$ au;
- the second dip is at 11.8 mA, corresponding to the crossing at $R = 7.8$ au;
- the third dip is at 5.5 mA, corresponding to the crossing at $R = 7.2$ au.

Below is the list of other prospective candidates for observing X-dips in spectral line profiles of He-like ions from laser-produced plasmas.
1. O VII He-gamma 17.78, perturber Li^{+3}, target lithium oxide (lithia) Li_2O (solid).
2. Si XIII He-beta 5.681 A, perturber Be^{+4}, target Be–Si binary alloy or beryllium-doped silicon.
3. Ca XIX He-beta 2.705 A, perturber C^{+6}, target calcium carbonate $CaCO_3$ (the common substance found in rocks and the main component of shells of marine organisms, snails, and egg shells).
4. V XXII He-beta 2.027 A, perturber N^{+7}, target vanadium nitride VN (solid).
5. Fe XXV He-beta 1.574 A, perturber O^{+8}, target iron oxides FeO, Fe_3O_4, Fe_2O_3 (crystalline solid).
6. Ca XIX He-gamma 2. 57 A, perturber F^{+9}, target calcium fluoride (fluorite) CaF_2 (mineral/solid).
7. Cu XXVIII He-beta 1.257 A, perturber F^{+9}, target copper fluoride CuF_2 (crystalline solid).

These results very significantly extend the range of fundamental data on CE between multicharged ions that can be obtained via the X-dip phenomenon, but not by any other method.

References

[1] Bransden B H and McDowell M R C 1992 *Charge Exchange and the Theory of Ion-Atom Collisions* (Oxford: Clarendon)
[2] Rosmej F B and Lisitsa V S 1998 *Phys. Lett.* A **244** 401
[3] Isler R C and Olson R E 1988 *Phys. Rev.* A **37** 3399
[4] Churilov S S, Dorokhin L A, Sidelnikov Y V, Koshelev K N, Schulz A and Ralchenko Y V 2000 *Contrib. Plasma Phys.* **40** 167
[5] Elton R C 1990 *X-Ray Lasers* (New York: Academic)
[6] Bunkin F I, Derzhiev V I and Yakovlenko S I 1981 *Sov. J. Quant. Electron.* **11** 981
[7] Vinogradov A V and Sobelman I I 1973 *Sov. Phys. JETP* **36** 1115
[8] Engel A, Koshelev K N, Sidelnikov Y V, Churilov S S, Gavrilescu C and Lebert R 1998 *Phys. Rev.* E **58** 7819

[9] Koshelev K N and Kunze H-J 1997 *Quant. Electron.* **27** 164

[10] Kunze H-J, Koshelev K N, Steden S, Uskov D and Wieschebrink H T 1994 *Phys. Lett.* A **193** 183

[11] Koshelev K N, Sidelnikov Y V, Churilov S S and Dorokhin L A 1994 *Phys. Lett.* A **191** 149

[12] Beiersdorfer P, Olson R E, Schweikhard L, Liebisch P, Brown G V, Crespo Lopez-Urrutia J, Harris C L, Neill P A, Utter S B and Widmann K 2000 *The Physics of Electronic and Atomic Collisions* ed Y Itikawa (New York: American Institute of Physics) p 626

[13] Boeddeker S, Kunze H-J and Oks E 1995 *Phys. Rev. Lett.* **75** 4740

[14] Sando K M, Doyle R O and Dalgarno A 1969 *Astrophys. J.* **157** L143

[15] Stewart J C, Peek J M and Cooper J 1973 *Astrophys. J.* **179** 983

[16] Queffelec J L and Girault M 1974 *C. R. Acad. Sci. Paris* **B279** 649

[17] Rang L Q and Voslamber D 1975 *J. Phys.* B **8** 331

[18] Preston R C 1977 *J. Phys.* B **10** 523

[19] Allard N and Kielkopf J 1982 *Rev. Mod. Phys.* **54** 1103

[20] Nelan E and Wegner G 1985 *Astrophys. J.* **289** L31

[21] Malnoult P, d'Etat B and Nguyen H 1989 *Phys. Rev.* A **40** 1983

[22] Leboucher-Dalimier E, Poquerusse A, Angelo P, Gharbi I and Derfoul H 1994 *J. Quant. Spectrosc. Radiat. Transf.* **51** 187

[23] Allard N F, Kielkopd J and Feautrier N 1998 *Astron. Astrophys.* **330** 782

[24] Allard N F, Drira I, Gerbaldi M, Kielkopf J and Spielfiedel A 1998 *Astron. Astrophys.* **335** 1124

[25] Oks E and Leboucher-Dalimier E 2000 *Phys. Rev.* E **62** R3067

[26] Oks E 1995 *Plasma Spectroscopy: The Influence of Microwave and Laser Fields. Springer Series on Atoms and Plasmas* vol 9 (New York: Springer)

[27] von Neumann J and Wigner E 1929 *Phys. Z.* **30** 467

[28] Gershtein S S and Krivchenkov V D 1961 *Sov. Phys. JETP* **13** 1044

[29] Kryukov N and Oks E 2012 *Phys. Rev.* A **85** 054503

[30] Power J D 1973 *Philos. Trans. R. Soc. Lond.* A **274** 663

[31] Komarov I V, Ponomarev L I and Slavyanov S Y 1976 *Spheroidal and Coulomb Spheroidal Functions* (Moscow: Nauka) [in Russian]

[32] Dalimier E, Oks E, Renner O and Schott R 2007 *J. Phys. B: At. Mol. Opt. Phys.* **40** 909

[33] Oks E 2015 *Breaking Paradigms in Atomic and Molecular Physics* (Singapore: World Scientific)

[34] Oks E 2006 *Stark Broadening of Hydrogen and Hydrogenlike Spectral Lines in Plasmas: The Physical Insight* (Oxford: Alpha Science International)

[35] Leboucher-Dalimier E, Oks E, Dufour E, Sauvan P, Angelo P, Schott R and Poquerusse A 2001 *Phys. Rev.* E **64** 065401

[36] Oks E and Leboucher-Dalimier E 2000 *J. Phys. B: At. Mol. Opt. Phys.* **33** 3795

[37] Djaoui A and Rose S J 1992 *J. Phys.* B **25** 2745

[38] Gauthier P, Rose S J, Sauvan P, Angelo P, Leboucher-Dalimier E, Calisti A and Talin B 1998 *Phys. Rev.* E **58** 942

[39] Sauvan P, Leboucher-Dalimier E, Angelo P, Derfoul H, Ceccotti T, Poquerusse A, Calisti A and Talin B 2000 *J. Quant. Spectrosc. Radiat. Transf.* **65** 511

[40] Talin B, Calisti A, Godbert L, Stamm R, Lee R W and Klein L 1995 *Phys. Rev.* A **51** 1918

[41] Renner O, Dalimier E, Liska R, Oks E and Šmíd M 2012 *J. Phys. Conf. Ser.* **397** 012017

[42] Dalimier E, Oks E and Renner O 2014 *Atoms* **2** 178

[43] Oks E *et al* 2014 *J. Phys. B: At. Mol. Opt. Phys.* **47** 221001

[44] Boldarev A S, Gasilov V A, Faenov A Y, Fukuda Y and Yamakawa K 2006 *Rev. Sci. Instrum.* **77** 083112

[45] Faenov A Y *et al* 1994 *Phys. Scr.* **50** 333

[46] Blasco F, Stenz C, Salin F, Faenov A Y, Magunov A I, Pikuz T A and Skobelev I Y 2001 *Rev. Sci. Instrum.* **72** 1956

[47] Oks E, Böddeker S and Kunze H-J 1991 *Phys. Rev.* A **44** 8338

[48] Sholin G V 1969 *Opt. Spectrosc.* **26** 275

[49] Demura A V and Sholin G V 1975 *J. Quant. Spectrosc. Radiat. Transf.* **15** 881

[50] Jackson E A 1967 *Phys. Rev.* **153** 235

[51] Liu C S 1976 *Advances in Plasma Physics* vol 6 ed A Simon and W B Thompson (New York: Wiley)

[52] Liu C S and Rosenbluth M N 1976 *Phys. Fluids* **19** 967

[53] Baldis H A, Villeneuve D M and Walsh C J 1986 *Can. J. Phys.* **64** 961

[54] Dalimier E and Oks E 2014 *J. Phys. B: At. Mol. Opt. Phys.* **47** 105001

[55] Demkov J N 1978 *Trudy Gosudarst. Opt. Inst. (Proc. State Opt. Inst.)* **43** 71 (in Russian)

[56] Nikitin S I and Ostrovsky V N 1976 *J. Phys. B: At. Mol. Opt. Phys.* **9** 3141

[57] Nadezhdin B B and Oks E 1990 *Opt. Spectrosc.* **68** 12

[58] Martin W C 1981 *Phys. Scr.* **24** 725

IOP Publishing

Advances in X-Ray Spectroscopy of Laser Plasmas

Eugene Oks

Chapter 3

Spectroscopic diagnostics of non-relativistic laser–plasma interactions

3.1 Brief theoretical introduction to the intra-Stark spectroscopy

Laser–plasma interactions can cause a variety of nonlinear phenomena in plasmas even when the laser intensity corresponds to the 'non-relativistic' situation. This is the range of laser intensities $I < I_{\text{threshold}}$, where $I_{\text{cr}} \sim 10^{18}$ W cm^{-2}. The phrase 'non-relativistic' in this context means that under such laser intensities, plasma electrons have non-relativistic velocities. In this situation the surface of the critical electron density N_{cr} (i.e. the surface enclosing the plasma volume where the laser radiation cannot penetrate) corresponds to

$$\omega = \omega_{\text{pe}}, \tag{3.1}$$

where ω is the laser frequency and ω_{pe} is the plasma electron frequency:

$$\omega_{\text{pe}} = (4\pi e^2 N_e/m_e)^{1/2}. \tag{3.2}$$

Here e and m_e are the electron charge and mass, respectively; N_e is the electron density. From equations (3.1) and (3.2) follows that in the non-relativistic situation, the critical density relates to the laser frequency as follows:

$$N_{\text{cr}} = m_e\omega^2/(4\pi e^2). \tag{3.3}$$

An advanced spectroscopic diagnostic of laser–plasma interactions is based on the intra-Stark spectroscopy, introduced in papers [1, 2] (see also book [3]). The name 'intra-Stark spectroscopy' is due to the physical analogy of this area with the intra-Doppler spectroscopy [4, 5]. It manifests as Langmuir-wave-caused dips/depressions (hereafter, L-dips) at certain locations of hydrogenic spectral line profiles. The L-dips were found experimentally as early as in 1977 [6]. The underlying theory was developed in papers [1, 6–8] for plasmas of relatively low electron densities and then generalized for high density plasmas in paper [2].

The intra-Stark spectroscopy has a rich underlying physics. In its basis there is an interplay/coupling of a quasimonochromatic dynamic electric field (e.g. of Langmuir waves) with a quasistatic electric field in plasmas, the latter being due to a low-frequency electrostatic plasma turbulence and/or the quasistatic part of the ion microfield. While the dynamic electric field producing the L-dips is quasimonochromatic, it causes a nonlinear dynamic-resonance effect of essentially *multifrequency* nature [8]. This is counterintuitive.

The resonance condition is the following

$$\omega_F = s\omega_{\mathrm{pe}}(N_e),\tag{3.4}$$

where ω_F is the separation between the Stark sublevels of a hydrogenic atom/ion due to the quasistatic electric field F and s is the number of the Langmuir plasmons involved in the resonance.

What is called for brevity L-dip, is actually a structure consisting of the primary local minimum and two surrounding bumps. This theoretical expectation was confirmed experimentally at the gas-liner pinch [2]. Figure 3.1 shows such experimental structure in the wing of the hydrogen Ly-alpha line (solid line). The dash-dotted line is the profile that would be at the absence of the Langmuir waves. This structure being superimposed with an inclined part of the line profile, could lead to a secondary minimum, as seen in figure 3.1 (or to a shoulder).

The location of the primary minimum is what the theory predicts and it used as one of the diagnostic tools. The location of the secondary minimum or shoulder is of no practical importance.

In paper [2] there were provided analytical results for the locations of the L-dips in any hydrogenic spectral line, i.e. in lines of any spectral series—for the situation where the quasistatic electric field in plasmas is dominated by the ion microfield. Here we reproduce only the locations of the L-dips $\Delta\lambda_{\pm}^{\mathrm{dip}}$ (counted from the unperturbed wavelength λ_0) in the profiles of the Ly-lines—because only Ly-lines (of multicharged hydrogenic ions) are typically observed in laser-produced plasmas:

$$\Delta\lambda_{\pm}^{\mathrm{dip}} = -\,[\lambda_0^2/(2\pi c)]\{sq\omega_{\mathrm{pe}} + [2(s\omega_{\mathrm{pe}})^3/(27n^3 Z_r Z_p w_{\mathrm{at}})]^{1/2} \\ \times [n^2(n^2 - 6q^2 - 1) + 12n^2 q^2 \pm 6n^2 q]\}.\tag{3.5}$$

Here $\omega_{\mathrm{at}} = m_e e^4/\hbar^3 \cong 4.14 \times 10^{16}\,\mathrm{s}^{-1}$ is the atomic unit of frequency, Z_r is the nuclear charge of the radiating atom/ion, Z_p is the charge of the perturbing ions; n and q are, respectively, the principal and electric quantum numbers of the upper energy level, from which the Lyman line originates ($q = n_1 - n_2$, where n_1 and n_2 are the parabolic quantum numbers). We recall that s is the number of the Langmuir plasmons involved in the resonance (3.4).

The first term in braces is due to the dipole interaction with the ion microfield. The second term in braces is a relatively small correction allowing for the spatial nonuniformity of the ion microfield. The second term takes into account the quadrupole interaction with the ion microfield.

It is important to emphasize that in the second term in braces in equation (3.5) is present only in the situation where the quasistatic electric field in plasmas is

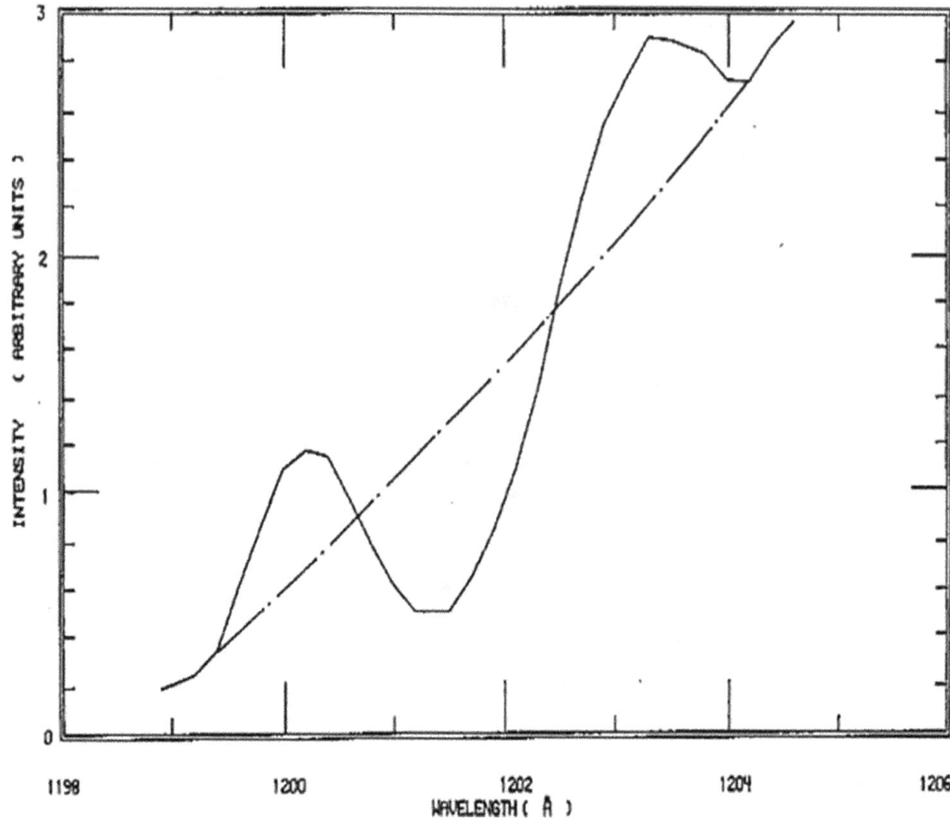

Figure 3.1. Bump–dip–bump structure (L-dip) in the wing of the experimental profile of the hydrogen Ly-alpha line observed at the gas-liner pinch [2] (solid line). The dash-dotted line is the profile that would be at the absence of the Langmuir waves.

dominated by the ion microfield. In the opposite situation, where the quasistatic electric field in plasmas is dominated by the low-frequency electrostatic plasma turbulence, the second term in braces in equation (3.5) is absent. This distinction is practically important: it allows us to find out experimentally whether a low-frequency electrostatic plasma turbulence was developed in the plasma under consideration—in addition to the Langmuir waves.

In summary, the practical importance of the intra-Stark spectroscopy is three-fold. First, this is the only one spectroscopic method for measuring the amplitude E_0 of the electric field of Langmuir waves. The amplitude E_0 can be deduced from the width of the experimental L-dips.

Second, the intra-Stark spectroscopy allows us to determine whether a low-frequency electrostatic plasma turbulence was also present along with the Langmuir waves. The simultaneous presence of both types of waves makes it possible to trace their origin to a specific plasma instability caused by the laser–plasma interaction.

Third, the intra-Stark spectroscopy allows measuring the electron density with a high accuracy. Namely, despite the intra-Stark spectroscopy being a passive method,

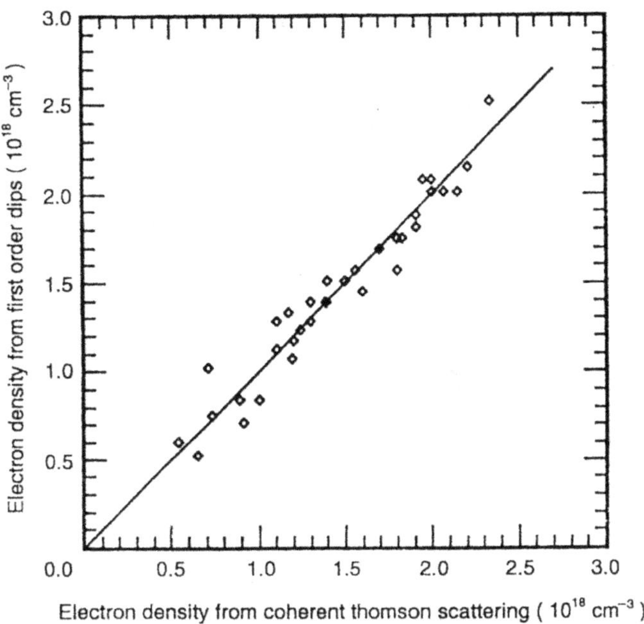

Figure 3.2. Comparison of the electron density deduced from the L-dips in the experimental profile of the hydrogen Ly-alpha line with the electron density determined by coherent Thomson scattering [2].

it allows measuring the electron density with the same high accuracy as coherent Thomson scattering, the latter being a much more complicated diagnostic to implement. This was shown in experiments at the gas-liner pinch [2] and illustrated in figure 3.2. It is seen that both methods yield the same high accuracy.

3.2 The first experimental implementation the intra-Stark-spectroscopy-based diagnostic in laser-produced plasmas

In paper [9] the authors presented the results of the experiment performed at the nanosecond Nd:glass laser facility in France—at the Laboratoire pour l'Utilisation des Lasers Intenses (LULI). The laser intensity was 2×10^{14} W cm^{-2}. The target was structured. The central part of it was made out of the (20–60) mm-thick aluminum (or powdered aluminum carbide Al_4C_3). This central part of the target was placed between plastic or magnesium substrates of the thickness 20–40 mm.

The electron density N_e and temperature T_e (close to the target surface) were initially estimated from hydrodynamic simulations. They yielded $N_e \sim 5 \times 10^{22}$ cm^{-3} and $T_e \sim 300$ eV.

The observed spectral line was Al XIII Ly-gamma. The primary experimental result was the same localized structures in the profiles of this spectral line. The structures were reproducible. The analysis showed that these structures are the L-dips.

Figure 3.3 shows the L-dips in the experimental profiles of the Ly-gamma line of Al XIII observed at three different distances from the target surface, corresponding to the electron densities 1.4×10^{22} cm^{-3}, $0.9 \times 4 \times 10^{22}$ cm^{-3}, and $0.5 \times 4 \times 10^{22}$ cm^{-3}.

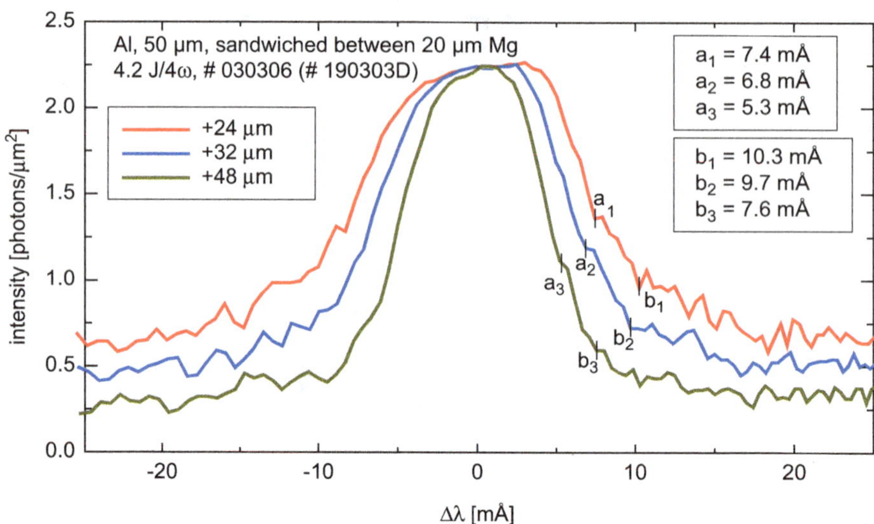

Figure 3.3. The L-dips in the experimental profiles of the Ly-gamma line of Al XIII [9]. Target: Al strip sandwiched between Mg substrate. The profiles were observed at three different distances from the target. The L-dips are marked by letters a and b in each experimental profile. Reprinted from [9], copyright 2006, with permission from Elsevier.

The electron densities were determined from the broadening of the experimental line profiles.

Then, by using the electron densities determined from the broadening, the authors of paper [9] marked in each experimental profile the locations of the possible L-dips, calculated for the known electron density. It turned out that each theoretical mark coincides with the observed local depression in the line profiles.

Specifically, each experimental profile exhibits two L-dips—marked by letters a and b—in the red wing. The possible L-dips in the blue wing merged with the noise.

Figure 3.4 shows L-dips in the experimental profiles of the Ly-gamma line of Al XIII observed while using a different target: Al_4C_3 target sandwiched between CH substrates [9]. The profiles were observed at four different distances from the target, corresponding to the electron densities 3.0×10^{22} cm^{-3}, $2.2 \times 4 \times 10^{22}$ cm^{-3}, $1.1 \times 4 \times 10^{22}$ cm^{-3}, and $0.5 \times 4 \times 10^{22}$ cm^{-3}. The electron densities were determined from the broadening of the experimental line profiles.

The L-dips are marked by letters a and b in each experimental profile. Vertical dashed lines mark positions of the charge-exchange-caused dips (X-dips)—see chapter 2 of this book. In the case presented in figure 3.4, the X-dips originates from the avoided crossings of the terms of the quasimolecule, consisting of Al^{12+} radiators perturbed by the ions C^{6+}.

The positions of the X-dips are practically independent of the electron density. Therefore, their positions are practically the same in all four experimental profiles— despite the four profiles corresponding to three different electron densities.

In section 3.1 it was noted that generally the L-dip is a bump–dip–bump structure, as, e.g. in figure 3.1. If two bump–dip–bump structures are located next to each

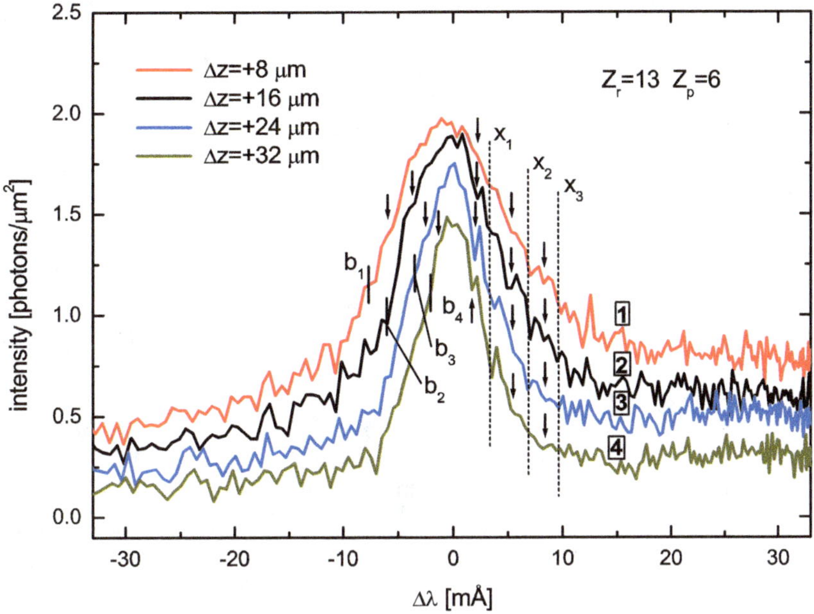

Figure 3.4. The L-dips in the experimental profiles of the Ly-gamma line of Al XIII [9]. Target: Al_4C_3 target sandwiched between CH substrates. The profiles were observed at four different distances from the target. The L-dips are marked by letters *a* and *b* in each experimental profile. Vertical dashed lines mark positions of the charge-exchange-caused dips (X-dips). The positions of the X-dips are practically independent of the electron density. Therefore, their positions are the same in all four experimental profiles—despite the three profiles correspond to four different electron densities. Reprinted from [9], copyright 2006, with permission from Elsevier.

other, this could produce four local minima of the intensity and so on. So, in figure 3.4 there are vertical arrows marking the secondary local minima for bump-dip-bump structures associated with the L- and X-dips.

It should be emphasized that if the X-dips in figure 3.4 were noise, then the corresponding part of the profiles in figure 3.3 would have looked as 'noisy' as in figure 3.4. However, the experimental profiles shown in figure 3.3 are much smoother than those from figure 3.4. This is consistent with the fact that for the case of Al strips sandwiched between Mg substrates there cannot be any X-dips: they exist only for quasimolecules Z_r–e–Z_p, where the nuclear charges Z_r and Z_p differ from each other.

In the experiment [9] the authors employed a vertical-geometry Johann spectrometer that recorded *simultaneously* two sets of spatially resolved spectra symmetrically located with respect to the maximum central wavelength. Figure 3.5 shows both of the symmetrical displaced line profiles of the experimental profile from figure 3.4, denoted as 2 in figure 3.2. The detailed superposition of these two records (individual measured points of the line profiles are shown in the insert) clearly demonstrates the reproducibility of the structures in both parts of the spectra (marked by vertical segments or by arrows) and thus the reliability of the interpretation of the experimental spectra.

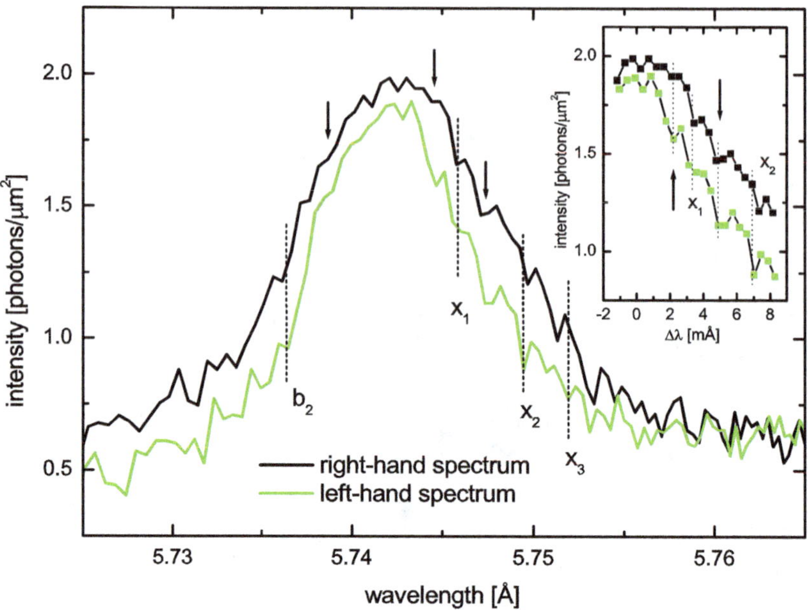

Figure 3.5. The experimental profile 2 from figure 3.4 and the spectrum reconstructed from the mirror-symmetric record on the same film [9]. Superposition of these two spectra demonstrates the reproducibility and reliability of the experimental data. Reprinted from [9], copyright 2006, with permission from Elsevier.

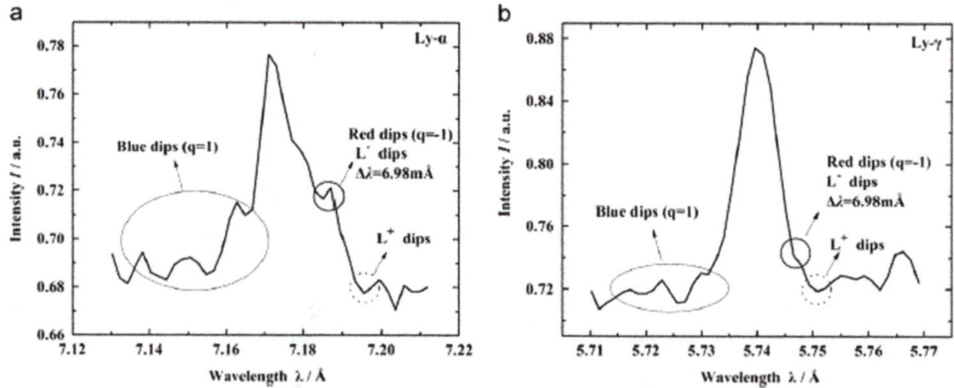

Figure 3.6. The experimental bump–dip–bump structures (L-dips) in the experimental profiles of the Al XIII Ly-alpha and Al XIII Ly-gamma lines, observed at a Z-pinch facility [10]. The L-dips in the blue wing of both experimental profiles merge in the noise. Reprinted from [10], copyright 2013, with permission from Elsevier.

The employment of the intra-Stark spectroscopy in super-high-density plasmas of multicharged ions is not limited to laser-produced plasmas. In paper [10] the authors presented a study of aluminum plasmas in a Z-pinch of the electron density $N_e = 5 \times 10^{21}$ cm^{-3}, and of the electron temperature $T_e = 500$ eV, and of the ion temperature $T_i = 10$ keV. The plasma was produced by imploding aluminum wire-array.

Figure 3.6 presents the experimental profiles of the Al XIII Ly-alpha and Al XIII Ly-gamma lines observed in paper [10]. The experimental profiles show bump–dip–bump structures typical for the resonance between the electric field of the Langmuir waves and the separation of Stark sublevels of Al XIII caused by the quasistatic electric field in the plasma—the resonance responsible for the L-dip phenomenon.

The electron densities derived from the L-dips in the red part of the experimental profiles were compared to the electron densities deduced from measurements of the plasma polarization shift of the entire spectral line (concerning the plasma polarization shift see, e.g. papers [11, 12] and references therein). There was only a 5% difference in electron density obtained by the two methods. This small difference was due to the uncertainty of determining the average electron temperature T_e required for the plasma polarization shift method (but not required by the method based on the L-dips).

3.3 Brief theory of satellites of dipole-forbidden spectral lines of He, Li, and the corresponding ions

The corresponding theoretical studies started from Baranger and Mozer paper [13]. They demonstrated that under a quasimonochromatic electric field (QEF) of the frequency ω, two satellites can appear at the frequencies $\omega_{\mathrm{sat}\pm} = \omega_{10} \pm \omega$. Here ω_{10} is the frequency of the dipole-forbidden spectral line, i.e. the frequency where it would appear if the electric field was static.

Baranger and Mozer considered the QEF to be three-dimensional. Later Cooper and Ringler [14] obtained more specific results for the case of a linearly polarized single-mode QEF.

The theoretical results of papers [13, 14] were obtained by the standard perturbation theory and therefore were valid only for the situation where the QEF is relatively weak. Oks and Gavrilenko [15] developed the analytical theory for stronger QEFs. The corresponding analytical solution from paper [15] was based on the adiabatic perturbation theory. In addition, the polarization of the satellites was analyzed in paper [15].

Later analytical results were obtained for the situation where the QEFs are very strong. These super-strong QEFs produce new features compared to weak QEFs. The first distinction is the following. While relatively weak QEFs produce only two satellites around the forbidden line and no satellites around the allowed line, the super-strong QEFs produce multi-satellite structures—both around the forbidden line and around the allowed line. The second distinction is that the multi-satellite structure around the forbidden line is characterized by a significant asymmetry of intensities. The corresponding analytical results can be found in book [3].

Finally it should be noted that magnetic interactions, such as spin–orbit coupling and spin-spin coupling, can significantly affect intensities of satellites—especially in the case of He-like and Li-like ions of a relatively large nuclear charge Z. Details can be found in book [3].

3.4 Laser-produced plasmas: spectroscopic diagnostic of the plasma interaction with an external oscillatory field

In this section we follow paper [16]. In the first part of paper [16] there were presented general principles of spectroscopic diagnostics of plasmas containing QEF. However, below we focus at the experimental results from paper [16] and their theoretical interpretation provided in that paper.

The experimental data needed for comparison with the theory were collected at the Jena multi-terawatt Ti:sapphire laser system JETI [17]. The laser routinely provides 1 J of energy, a pulse duration of 80 fs (FWHM), and repetition rate of 10 Hz. In the experiment, the laser pulse with the wavelength of 0.8 μm was stretched to 12.5 ps and the energy of 0.65 J delivered at the entrance window of the interaction chamber was split into two beams [18]. The near-target configuration of these beams, which were focused by off-axis parabolic mirrors to foci with a diameter of about 20 μm, is schematically shown in figure 3.7. The plasma producing laser beam $(0.2 \text{ J}, 5 \times 10^{15} \text{ W cm}^{-2})$ was introduced to the chamber via a delay line and focused to a flat tip of the tapered Al target. The second beam $(0.45 \text{ J}, 1.2 \times 10^{16} \text{ W cm}^{-2})$ with the axis parallel to the target surface was hitting the plasma plume transversally. The axis of the plasma expansion, the electric field of the second beam and the direction of the x-ray spectra observation were mutually perpendicular. The electric field vector of the second laser beam was approximately perpendicular to the line of sight of the spectra, i.e. the transverse-field-affected line profiles were observed.

Both focused laser beams were spatially overlapped and temporally synchronized with precision better than 1 ps. The timing and positioning of the perturbing beam above the target surface with respect to the plasma critical density surface (for the laser wavelength of 0.8 μm, this corresponds to the electron density $N_c = 1.7 \times 10^{21} \text{ cm}^{-3}$) and the duration of the Al Heβ emission were optimized using predictions of one-dimensional plasma simulations [18]. The conclusions following from this modeling can be recapitulated in two points. Even at a short distance of 20 μm above the target, the Al Heβ emission starts safely after termination of the plasma-creating beam that precludes direct field effects of this beam on the observed line emission.

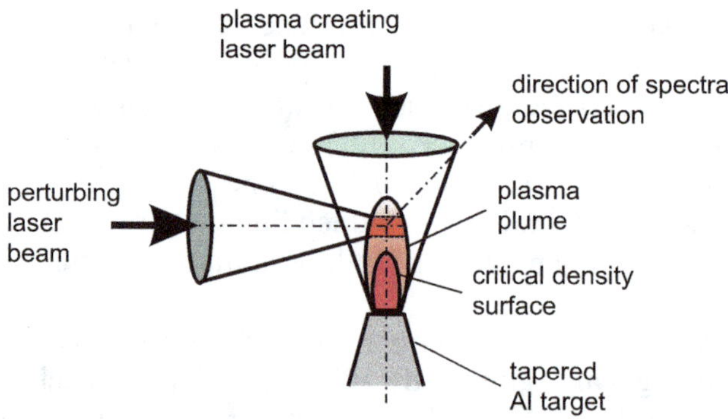

Figure 3.7. Schematic diagram of laser beams and target geometry in the experiment [16].

In contrast, a good temporal overlap of the plasma Al Heβ emission with the 50 ps-delayed perturbing beam is favorable for observation of field effects in the spectral line profiles. The macroscopic plasma parameters simulated in more detail by two-dimensional Lagrangian hydro-code CHIC [19] indicate that at the distance of 20 μm from the target, the quasi-flat distributions of the electron density N_e (well below the $N_c/4$) and temperature T_e (slightly above 120 eV) exist with a transverse dimension comparable to the focal spot diameter. These parameters are well compatible with the desired laser field penetration into the preformed plasma and with the observation of the spectral line Al Heβ emission.

The x-ray spectra were measured with the toroidally bent crystal spectrometer (TCS) using a crystal of quartz (10-1) with bending radii 150/106.4 mm and fitted with a CCD camera. The fulfillment of the focal condition for Al Heβ radiation [20] at the source-to-crystal distance of 83.8 mm resulted in a spectral resolution of 6100, a magnification of 1.77, and a spatial resolution of approximately 12 μm. The spectral range of 28 eV covered by the spectrometer was relatively small but sufficient to study the full profile of the Al Heβ emission centered at 1867.7 eV. Benefiting from the extremely high collection efficiency of the TCS, the spectra were recorded in single laser shots. Further details concerning the experimental configuration and the spectra calibration can be found in paper [18].

The unperturbed Al Heβ emission was observable up to a distance of about 90 μm above the target. The presence of the transverse laser beam introducing the external electric field into the plasma environment slightly increased the noise of the spectral records and, most importantly, induced distinct modulations into the line profiles. The field effects were varied by changing the time delay between both laser beams and the distance of the perturbing beam above the target. In agreement with simulations, the most pronounced spectra perturbations were observed with the transverse laser beam delayed by 50 ps and introduced at the distance of 20 μm above the target. The reproducibility of the perturbing-beam-induced structures in spectral profiles Al Heβ is demonstrated in figure 3.8, where the measured points are smoothed using the 5 point fast Fourier transform. Albeit not all fine details are resolved in each spectrum, the presence of the dominant extremes in spectral line profiles is reasonably well reproducible. The identification of these local maxima follows from detailed simulations presented below.

Now we present the Floquet–Liouville theory. The Liouville space, usually employed to deal with the calculation of Stark profiles in dense plasmas, and the Floquet theory, developed to solve time periodic problems, have been joined together to solve the time-dependent Liouville equation in a so-called Floquet–Liouville formalism [21]. In the following we summarize the main steps of this theory and briefly discuss the control of the accuracy for the convergence of the simulated profiles.

The starting point of the line shape calculation is the determination of the dipole correlation function

$$C(t) = \int Q(\mathbf{F}) \, \mathrm{Tr}_{ae}[\mathbf{d}\, U(\mathbf{F}, t)(\rho^{(a)}\rho^{(e)}\mathbf{d})] \, d^3\mathbf{F}. \tag{3.6}$$

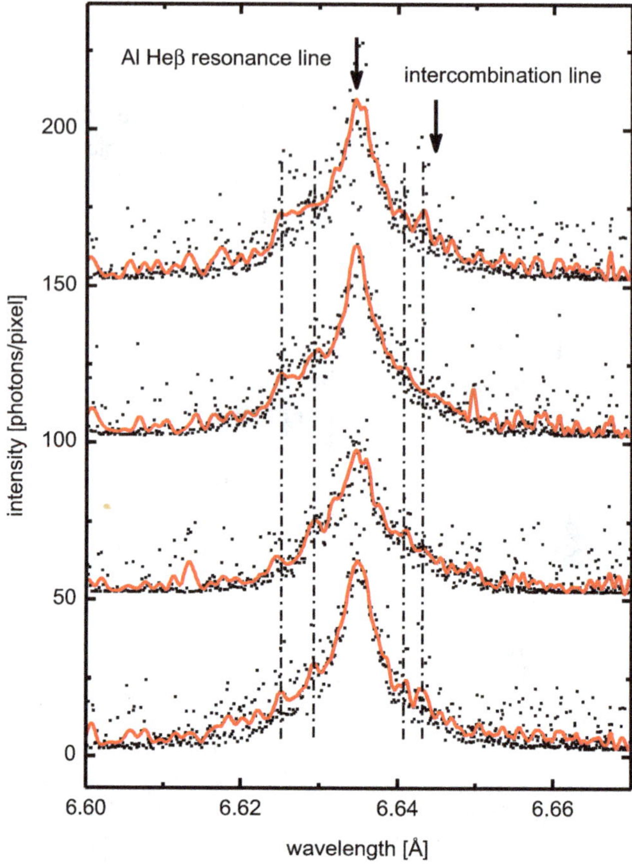

Figure 3.8. Reproducibility of Al Heβ line profiles recorded in different shots in the experiment [16]. The experimental data (dots) are smoothed using the 5 point FFT (solid line). The bottom profile in this figure is selected for comparison with theory and simulations.

In equation (3.6) $\rho^{(a)}$ and $\rho^{(e)}$ stand for the emitter and the free-electrons density matrix, respectively. The trace Tr_{ae} involves all emitters and free electron states; d is the electric dipole moment operator. The ions are considered as quasi-static during the time of interest, and the ionic micro-field $\mathbf{F}$ is then taken constant during the radiative transitions. Profiles of the same spectral line emitted by different radiators depend on the field $\mathbf{F}$ as a parameter. The final profile of the spectral line is obtained by averaging over the ionic micro-field distribution $Q(\mathbf{F})$.

The evolution operator $U(\mathbf{F},t)$ depends parametrically on $\mathbf{F}$. In the Liouville formalism, it is expressed as follows:

$$U(\mathbf{F},\, t) = \exp\left(-i\int_{o}^{t} L(\mathbf{F},\, t')dt'\right). \tag{3.7}$$

Here $L(\mathbf{F},t)$ is the Liouville operator, or more precisely, a sum of two Liouville operators: one operating on the emitter states $L_r(\mathbf{F},t)$ and the second operating only on free-electron states $L_e(t)$,

$$L(\mathbf{F}, t) = L_r(\mathbf{F}, t) + L_e(t), \tag{3.8}$$

with $L_r(\mathbf{F}, t) \equiv L_r(t) = L_a + L_i(\mathbf{F}) + L_{of}(t)$. Here L_a, L_i, L_{of}, L_e stand for the Liouville operators for the isolated emitters, the interaction of the emitter with the ion micro-field, the interaction of the emitter with the QEF, and the interaction of the emitter with the free electrons, respectively.

Following the assumption that the effect of QEF on the free-electron field is neglected, the trace over electron states of the commutator $[L_r, L_e]$ vanishes. Thus the trace in the correlation function in equation (3.6) acts on a product of exponentials, and equation (3.7) can be rewritten as

$$U(\mathbf{F}, t) = \exp\left(-i\int_o^t L_r(\mathbf{F}, t')dt' - \phi_e t\right) = \exp\left(-i\int_o^t (L_r(\mathbf{F}, t') - i\phi_e)dt'\right) \tag{3.9}$$

In equation (3.9) we introduced the time-independent electron broadening operator ϕ_e (under the impact approximation) and averaged over all initial times.

The Liouville operator $L(\mathbf{F},t)$, depending on the oscillating field, can be transformed into a time-independent Floquet–Liouville L_F. This new operator satisfies, as demonstrated in [21], the eigen-value equation involving *a matrix of infinite dimension*:

$$\sum_{\sigma\tau}\sum_{k}\langle\alpha\beta; n \mid L_F \mid \sigma\tau; k\rangle\langle\sigma\tau; k|\Omega_{\mu\nu,m}\rangle = \Omega_{\mu\nu,m}\langle\alpha\beta; n|\Omega_{\mu\nu,m}\rangle. \tag{3.10}$$

In equation (3.10) $|\Omega_{\mu\nu,m}\rangle$ represents the eigenvector corresponding to the eigen-value $\Omega_{\mu\nu,m}$, and the states $|\alpha\beta; n\rangle = |\alpha\beta\rangle \otimes |n\rangle$ are expressed on the generalized tetradic-Fourier basis. The time-independent operator L_F matrix elements can be written more explicitly as:

$$\langle\alpha\beta; n \mid L_F \mid \mu\nu; m\rangle = L_{\alpha\beta,\mu\nu}^{(m-n)} + m\omega\delta_{\alpha\mu}\delta_{\beta\nu}\delta_{nm}, \tag{3.11}$$

where

$$L_{\alpha\beta,\mu\nu}^{(k)} = H_{\alpha\mu}^{(k)}\delta_{\beta\nu} - H_{\beta\nu}^{(k)}\delta_{\alpha\mu} - i\phi_{\alpha\beta,\mu\nu}^{el}\delta_{k0}, \tag{3.12}$$

$$H_{\alpha\beta}^{(k)} = H_{\alpha\beta}\delta_{k,0} + V_{\alpha\beta}(\delta_{k,1} + \delta_{k,-1}), \tag{3.13}$$

$$H_{\alpha\beta} = E_\alpha\delta_{\alpha\beta} - \langle\alpha|\mathbf{d} \cdot \lessgtr\mathbf{F}|\beta\rangle, \tag{3.14}$$

$$V_{\alpha\beta} = -\frac{1}{2}\langle\alpha|\mathbf{d} \cdot \mathbf{E}|\beta\rangle. \tag{3.15}$$

In these expressions $H_{\alpha\beta}$ is the ionic Stark Hamiltonian, and $V_{\alpha\beta}$ the interaction with the QEF; $\mathbf{d}$ is the dipole moment of the emitter, and $\mathbf{F}$ and $\mathbf{E}$ are the ionic micro-field and the QEF, respectively.

The matrix (3.10) has periodic properties, i.e.

$$\Omega_{\mu\nu,m+k} = \Omega_{\mu\nu,m} + k\omega, \tag{3.16}$$

$$\langle \alpha\beta; n+k | \Omega_{\mu\nu,m+k} \rangle = \langle \alpha\beta; n | \Omega_{\mu\nu,m} \rangle. \tag{3.17}$$

Due to the non-Hermitian nature of L_F, there exists a bi-orthogonal relationship between left and right eigenvectors

$$\left\langle \Omega_{\mu\nu,m}^{-1} \middle| \Omega_{\alpha\beta,n} \right\rangle = \delta_{\alpha\mu}\delta_{\beta\nu}\delta_{nm}. \tag{3.18}$$

In the following we emphasize the key-points of the Floquet's formulation. The evolution operator related to the density operator ρ is expressed via two matrix, $\phi(t)$, a matrix of eigen-states of L_F, and a diagonal matrix Q, the elements of which are the quasi-energies $\Omega_{\alpha\beta,n}$ [21], i.e.

$$U(t, t_0) = \rho(t)\rho^{-1}(t_0), \tag{3.19}$$

$$\rho(t) = \phi(t)e^{-iQt}. \tag{3.20}$$

Taking into account the periodic property (3.17) and the bi-orthogonal relationship (3.18), the evolution operator can be expanded compactly as

$$U(t, t_0) = \sum_{\substack{\alpha\beta \\ n}} \sum_{\mu\nu} | \alpha\beta; n\rangle\langle\alpha\beta; n | \exp(-iL_F(t - t_0)) | \mu\nu; 0\rangle\langle\mu\nu; 0 |. \tag{3.21}$$

Consequently, its matrix elements are

$$U_{ab,a'b'}(t, t_0) = \sum_{n}\langle ab; n|\exp(-iL_F(t - t_0)) |a'b'; 0\rangle e^{in\omega_L t}, \tag{3.22}$$

where we have written

$$\langle ab|\alpha\beta; n\rangle = \langle ab|\alpha\beta\rangle \otimes | n\rangle \equiv \langle ab|\alpha\beta\rangle e^{in\omega_L t}. \tag{3.23}$$

In equation (3.22), the periodic property of the eigenvector allowed the removing of the sum over $|m\rangle$ states, and setting the state $|m\rangle = |0\rangle$, arbitrarily.

We see in equation (3.22) that the evolution operator is defined on the finite tetradic basis $|\alpha\beta\rangle$, and the Floquet–Liouville operator is defined on an infinite generalized tetradic Fourier basis $|\alpha\beta; n\rangle$. The reduction of the tetradic Fourier basis $|\alpha\beta; n\rangle$ on the tetradic basis $|\alpha\beta\rangle$ is performed by summing over all projections of the Floquet–Liouville operator on the subspace with a fixed mode n.

The evolution operator given by equation (3.9) involves an average over electron states and a time-independent electron broadening operator. Therefore, in accordance with paper [22], we need to average the evolution operator over all initial times

t_0, keeping constant the elapsed time $s = t - t_0$. For this purpose the time evolution operation is expressed in terms of t_0 and s as

$$U_{\alpha\beta,\mu\nu}(s, t_0) = \sum_n \langle \alpha\beta ; n \mid \exp(-iL_F s) \mid \mu\nu ; 0\rangle e^{in\omega_L s} e^{in\omega_L t_0}. \tag{3.24}$$

Since the only dependence on t_0 is in the exponential, the non-zero contribution for the time-average evolution operator $\bar{U}_{\alpha\beta,\mu\nu}(s)$ will be only for $n = 0$, i.e.

$$\bar{U}_{\alpha\beta,\mu\nu}(s) = \langle \alpha\beta; 0|\exp(-iL_F s)|\mu\nu; 0\rangle. \tag{3.25}$$

Finally, the average over initial times has removed the sum over Fourier modes, and only the mode $n = 0$ remains. Thus we have defined all necessary operators to calculate the spectral line profiles.

The presence of the QEF introduces a polarization of the space, while the ionic micro-field $\mathbf{F}$ may have any orientation with respect to the linearly-polarized QEF $\mathbf{E}$. We introduce the complex vector basis $e_0 = e_z$, $e_1 = -\frac{1}{\sqrt{2}}(e_x + ie_y)$, $e_{-1} = \frac{1}{\sqrt{2}}(e_x - ie_y)$, by choosing the z-axis along $\mathbf{E}$. The electric dipole moment operator d occurring in equation (3.8), the ionic micro-field $\mathbf{F}$ and the QEF $\mathbf{E}$ can be decomposed on the vector basis so that the scalar products in equations (3.14) and (3.15) will be expressed as $\mathbf{d} \cdot \mathbf{E} = d_0 E$ and $\mathbf{d} \cdot \mathbf{F} = d_0 F_0 - d_1 F_{-1} - d_{-1} F_1$. For the comparison with experimental results it is useful to introduce parallel and perpendicular components of $\mathbf{F}$ with respect to $\mathbf{E}$:

$$F_0 = F_{//}, \quad F_1 = -\frac{1}{\sqrt{2}}F_\perp e^{i\varphi}, \text{ and } F_{-1} = \frac{1}{\sqrt{2}}F_\perp e^{-i\varphi}. \tag{3.26}$$

Due to the rotation invariance of the z-axis, one can choose $\varphi = 0$. The correlation function $C_q(t)$ can be related to each polarization state distinguished by the suffix q.

Based on the above results, the intensity for each polarization q can be expressed as follows:

$$I_q(\omega) = \frac{1}{\pi} \text{Re} \int_0^\infty e^{i\omega t} \int Q(\mathbf{F}) \left\langle d_q^\dagger \middle| U(\mathbf{F}, t) \middle| \rho^{(a)} d_q \right\rangle d^3\mathbf{F} \, dt. \tag{3.27}$$

In accordance with paper [21], this can be re-written as:

$$I_q(\omega) = \int Q(\mathbf{F}) \left[\frac{1}{\pi} \text{Re} \int_0^\infty e^{i\omega t} \left\langle d_q^\dagger \middle| P_0 \exp(-iL_F t)P_0 \middle| \rho^{(a)} d_q \right\rangle dt \right] d^3\mathbf{F}, \tag{3.28}$$

where P_0 is the projector on the Floquet subspace $n = 0$

Following the procedure from paper [23] for the numerical treatment of the double summation involved in equation (3.28), we diagonalize the Floquet–Liouville operator $L_F(\mathbf{F})$ for every ionic field $\mathbf{F}$. The Fourier transform can be performed easily, leading to an infinite diagonal matrix [21] in which all Floquet modes n are considered. In the numerical calculation, the infinite dimension of this matrix has to be reduced to a finite number of modes n. It has been shown in paper [21] that the number of modes to be included in the calculation can be estimated for any desired accuracy. For the integration over the ionic micro-field, a discretization of both the micro-field intensity and the direction had to be done, involving Gauss–Legendre quadrature [21].

In summary, the Floquet–Liouville operator, introduced to solve the time-dependent problem (Stark effect in a time-dependent QEF), leads to a time-independent treatment carried out via an operator of the infinite dimension. Nevertheless, the periodic property of this operator allows one to consider only the projection of the Floquet–Liouville operator on the Floquet subspace n. Depending on the coupling values between the Floquet subspaces for the Floquet–Liouville operator, for any desired accuracy of the line shape calculation, there will exist a value of n, at which the Floquet subspace can be truncated.

In the experiment described in paper [16], the electric field of the additional laser beam was perpendicular to the line of sight, so that the observed profiles were affected by the transverse QEF. The profile selected for the comparison with simulations is the bottom profile in figure 3.8. The spatially-dependent macroscopic parameters, controlling the observed profile, are the electron density $N_e(x)$, the electron temperature $T_e(x)$ and the QEF $E(x)$. From 2D hydro-simulations results [18, 19], obtained 50 ps (which is the additional laser beam delay) after the maximum of the plasma-creating laser beam, $N_e(x)$ shows a plateau below the critical density ($N_c = 1.7 \times 10^{21}$ cm^{-3}) and T_e shows a smooth gradient in the range 120–150 eV. As for the QEF intensity E in vacuum, it is related to the intensity I of the additional laser by the well-known formula $E = (\frac{2I}{\varepsilon_0 c})^{1/2}$ in SI units, yielding to 3.01 × 10^9 V cm^{-1} for the laser intensity of interest 1.2 × 10^{16} W cm^{-2}. This value corresponds to $0.58 E_0$ where $E_0 = 5.146$ GV cm^{-1} is the atomic unit of the electric field.

However, laser–plasma interactions should significantly change the QEF strength in the plasma; besides, the QEF in the plasma would be characterized by strong gradients.

Let us first discuss the effect of the macroscopic parameters N_e, T_e and E on the spectra without performing spatial integrations. The spatial inhomogeneity of these parameters—especially of $E(x)$—will be taken into account at the subsequent step.

Al Heβ Floquet–Liouville simulations were performed for the plasma parameters of interest, N_e from 2.5 × 10^{20} cm^{-3} to 1.5 × 10^{21} cm^{-3}, $T_e = 150$ eV, and for varying QEF strengths $E = 0$, 0.1, 0.3 and 0.6 unit of E_0. The electronic temperature was chosen to be 150 eV for all simulations. The results are given in figures 3.9 and 3.10 for 'parallel' profiles (i.e. polarized along the QEF) and 'perpendicular' profiles (i.e. polarized perpendicular to the QEF), respectively.

From these simulations it is obvious that the variation of the density, in the above range, has only a minor effect on the spectra, while the variation of the QEF strength affects the spectra and is therefore the leading parameter. The variation of this parameter along the line of sight x is discussed in detail below: it will be the decisive factor for the positions and the intensities of the laser satellites.

The effects of laser–plasma interaction relevant to this experiment are of two kinds. First, there is Raman and Brillouin backscattering, due to which the QEF in the plasma can be significantly higher than the strength of the additional laser beam

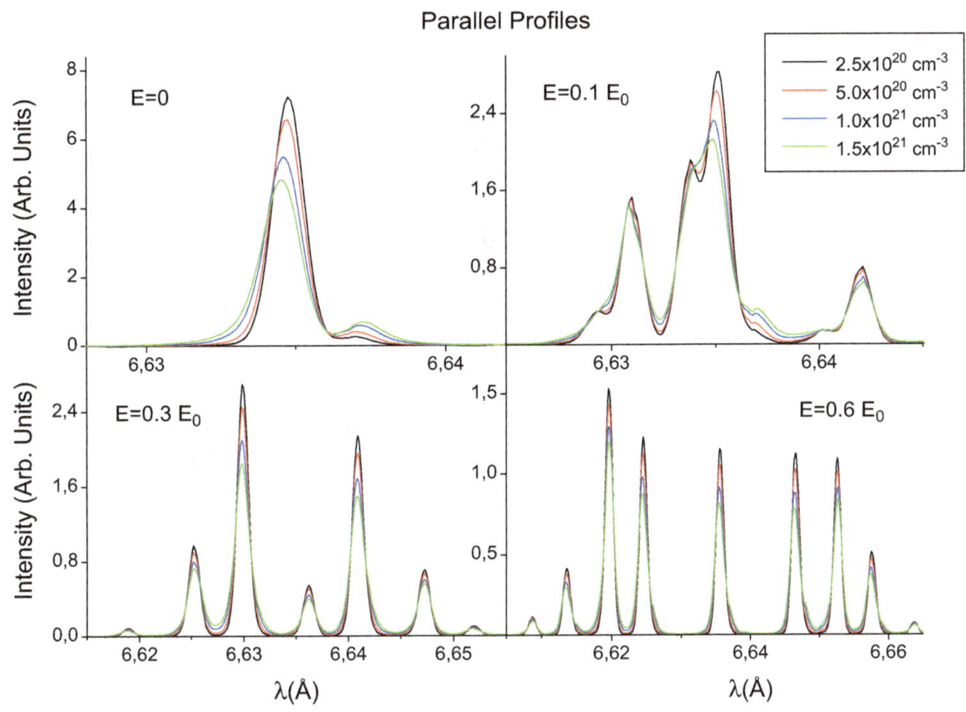

Figure 3.9. Floquet–Liouville simulations for the parallel Al Heβ profiles (i.e. polarized along the QEF) for $N_e = 2.5 \times 10^{20}$ cm^{-3} to 1.5×10^{21} cm^{-3}, $T_e = 150$ eV, and QEF strengths $E = 0, 0.1, 0.3$ and 0.6 of $E_0 = 5.146$ GV cm^{-1}, which is the atomic unit of the electric field [16].

in vacuum. Second, strong plasma fluctuations are excited due to the parametric instability, resulting in the spatial modulation of the electron density.

The simulations show that the spatial distribution of the electric field depends on the initial electron density profile in the plasma crossed by the second laser. Parabolic electron density profiles were introduced with different maximum densities.

The physical phenomena involved are sensitive to whether the maximum electron density is below or above one quarter of the critical density N_c. Above $N_c/4$, under the conditions of the present experiment, the dominant process is the stimulated Brillouin backscattering in the so-called strong coupling regime leading to the formation of transient phenomena, such as plasma cavities and transverse electromagnetic solitons [24, 25].

Figure 3.11 presents the results of the kinetic PIC simulations of the spatial distribution of the transverse time-averaged electric field E in the plasma for electron densities $0.1N_c$, $0.2N_c$, and $0.3N_c$. It shows significant changes compared to the laser field E_0 in vacuum. The changes become especially dramatic for $N_e > N_c/4$, as seen at the right part of figure 3.11, corresponding to $N_e = 0.3N_c$.

For a detailed comparison with the observed Al Heβ spectra emitted from plasmas in the presence of QEF, we introduce synthetic line profile simulations

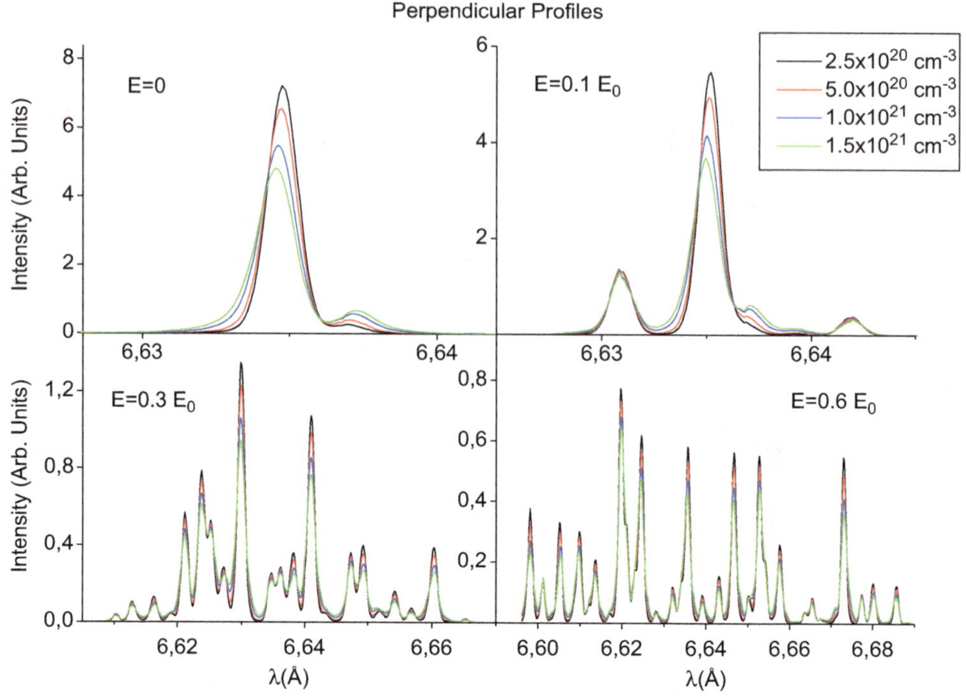

Figure 3.10. Floquet–Liouville simulations for the perpendicular Al Heβ profiles (i.e. polarized perpendicular to the QEF) for $N_e = 2.5 \times 10^{20}$ cm^{-3} to 1.5×10^{21} cm^{-3}, $T_e = 150$ eV, and QEF strengths $E = 0$, 0.1, 0.3 and 0.6 of $E_0 = 5.146$ GV cm^{-1}, which is the atomic unit of the electric field [16].

obtained within the Floquet–Liouville formalism using electric field PIC simulations. The spectroscopic diagnostic has no time resolution. So we assume that the synthetic profile is a sum of two terms. The first term is due to the Al Heβ emission from the *perturbed plasma* existing only at the time interval T corresponding to the duration of the second laser pulse. The second term is due to the total Al Heβ emission of the duration Δt_{unpert} from the *unperturbed plasma* before the second laser pulse. Both terms involve space integrations along the line of sight x and thus take into account the spatial inhomogeneities.

Hence the synthetic profile is expressed as following:

$$\Phi_\nu = \int_0^T \left[1 - \exp\left(-\int_0^L k(\bar{N}_e, \bar{T}_e) \; \varphi_\nu(E(x, t), \bar{N}_e, \bar{T}_e) \, dx \right) \right] dt$$

$$+ \Delta t_{\text{unpert}} \left[1 - \exp\left(-\int_0^L k(N_e^0, \bar{T}_e) \; \varphi_\nu^0(N_e^0, \bar{T}_e) \, dx \right) \right] \tag{3.29}$$

In the '*unperturbed profile*' contribution, the normalized profile ϕ_ν^0 and the absorption coefficient k are calculated for a time/space average density N_e^0 characterizing the plasma before the second laser pulse. In the '*perturbed profile*' contribution, the normalized profile ϕ_ν and the absorption coefficient k are calculated for a time/

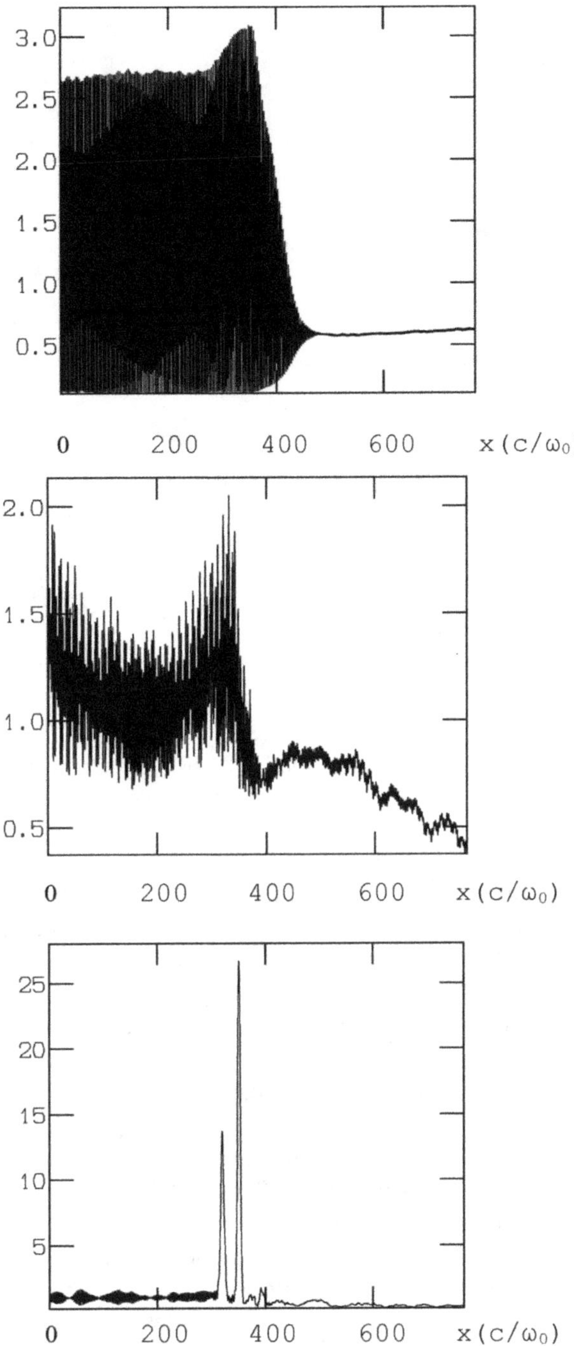

Figure 3.11. Spatial distribution of the energy density E^2/E_0^2 of the transverse quasi-monochromatic electric field inside the plasma (the pulse is coming from the left) [16]. The values are normalized to the energy density of the incident laser field. For these PIC simulations, the electron density vary from top to bottom: $0.1N_c$, $0.2N_c$, $0.3N_c$, where $N_c = 1.7 \times 10^{21}$ cm^{-3} is the critical density. The abscissa scale is in units c/ω_0 where ω_0 is the unperturbed frequency of the second laser and c is the speed of light.

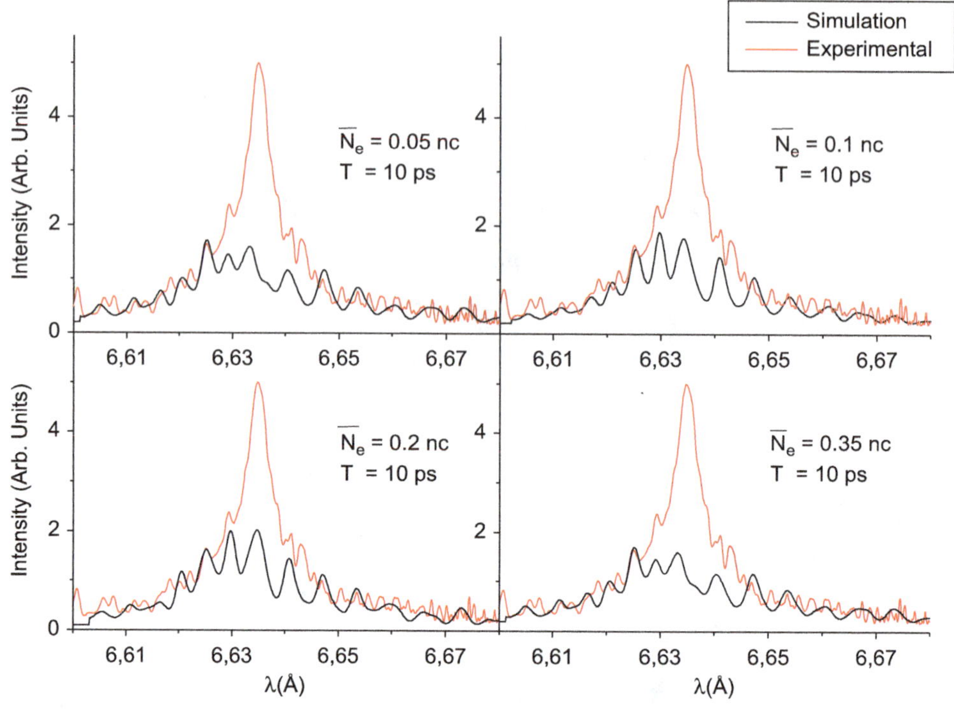

Figure 3.12. Comparison between the simulated 'perturbed' Al Heβ profiles (i.e. emitted only by the plasma perturbed by the second laser) and the experimental profile for different initial electron densities $\bar{N}_e$ [16].

space average density $\bar{N}_e$. This assumption is reasonable because the density affects the profiles weakly compared to the effect of the QEF. The average density $\bar{N}_e$ is smaller than N_e^0. This is because by the time the pulse of the second laser reaches its maximum, the plasma has already expanded transversally compared to its dimension during the emission of the 'unperturbed profile'.

So, the temporal and spatial dependences of the electric field are taken into account in the integrations in equation (3.29). These dependences are related to the transverse inhomogeneity of the electron density.

For all simulations, we chose an average temperature $\bar{T}_e = 150$ eV (this parameter is not critical), the optical depth $L = 50$ μm, the average electron density $N_e^0 = 0.6N_c$, and the second laser pulse duration $T = 10$ ps (FWHM). Finally the synthesized profiles were convoluted with a Gaussian instrument function of 0.3 eV FWHM.

Figure 3.12 presents simulated perturbed profiles for the initial electron densities $0.05N_c$, $0.1N_c$, $0.2N_c$, and $0.35N_c$ (these densities are chosen also for the average densities $\bar{N}_e$ of the perturbed profiles). These profiles are compared to the experimental profile. The crucial points in the comparison are the intensities, the broadenings, and the positions of the satellites located in the near wings of Al Heβ.

Figure 3.13 shows the full Al Heβ profile (i.e. both terms in equation (3.29)), simulated for $\bar{N}_e = 0.2N_c = 3.3 \times 10^{20}$ cm^{-3} and $\Delta t_{\text{unpert}} = 8.9$ ps, and the comparison with the experimental profile. One of the essential points in the final comparison is a

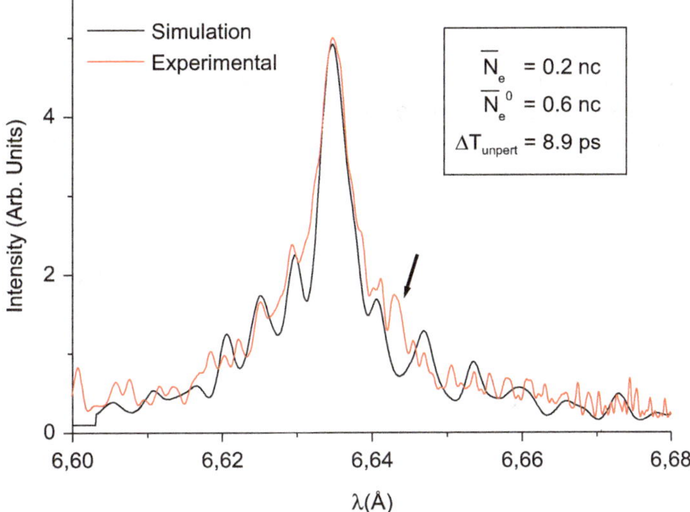

Figure 3.13. Comparison between the simulated total ('perturbed' + 'unperturbed') Al Heβ profile and the experimental profile [16]. The parameters shown in the frame correspond to the best fit.

good fit with the intensity and the broadening of the central part of the Al Heβ line, as well as with the first two satellites in the blue side and the first two satellites in the red side. More distant satellites of the simulated profile are located in the far wings, where the experimental profile merges into the noise.

In the red part of the experimental profile, between the peaks identified as the first and the second red satellites, there is an additional local maximum denoted by an arrow in figure 3.13. Its position agrees with the dipole forbidden transition $1s3s\,^1S_0$–$1s^2\,^1S_0$. Its emergence can be explained by the coupling of singlet $^1S_0$ and $^1D_2$ states via the quadrupole interaction with the ion micro-field. This interaction, being due to the intrinsic inhomogeneity of the ion micro-field, can significantly enhance some dipole forbidden spectral lines, as was first shown by Gaisinsky and Oks [26]. Detailed simulation and interpretation of this phenomenon has been published in paper [27].

For the 'perturbed' profile, the quadratic Stark shift $\Delta\lambda_{qss}$ by the QEF was taken into account: $[\Delta\lambda_{qss},\ mA] = 0.2\ [E(x),\ GV\ cm^{-1}]^2$. Intensities of different satellites depend in different ways on $E(x)$. Therefore, after the spatial integration each satellite has its individual quadratic Stark shift to the red side. In other words, the quadratic Stark shift affects the separations between the satellites, so that its inclusion was essential for the adequate interpretation of the experimental profile. We note that even at the vacuum value of the laser field $E_0 = 3\ GV\ cm^{-1}$, the quadratic Stark shift is quite significant: 1.8 mA. It is even more significant in the plasma, where the QEF is greater than the laser field in vacuum. We also recall that the quadratic Stark effect due to the quasi-static part of the ion micro-field $\mathbf{F}$ is negligible because the effective value of $\mathbf{F}$ at $N_e = 3.3 \times 10^{20}\ cm^{-3}$ is about three orders of magnitude smaller than the QEF.

Also, the 'perturbed' profile had to be shifted by 1.2 mA to the blue side with respect to the 'unperturbed' profile. Most probably this is the Doppler shift caused by the additional transverse expansion of the plasma under the second laser. After the second laser hits the plasma, the temperature T increases, leading also to a rise of the charge state Z_i of plasma ions. Then the edge of the plasma closest to the observer expands into the vacuum toward the observer with an increased velocity. The expansion velocity should be of the order of the ionic sound velocity $c_s = (3Z_iT_e/M_i)^{1/2}$, where $T_e = T$ and M_i is the mass of the plasma ions. The combined effect of the increases of T_e and Z_i translates in the rise of the expansion velocity of the perturbed plasma by a factor of two compared to the expansion velocity of the unperturbed plasma [28]. Thus, from the PIC simulations [18], the relative difference of the expansion velocities can be estimated as $(6{-}8) \times 10^6$ cm s^{-1}. This is consistent with the interpretation of the above relative shift of the perturbed and unperturbed profiles, because the shift of 1.2 mA corresponds to the difference in expansion velocities of 5.4×10^6 cm s^{-1}.

Indeed, all other sources of the shift are much smaller. We estimated the electron oscillatory shift (EOS) $\Delta\lambda_{\mathrm{EOS}}$ [3, 29, 30] and the plasma polarization shift $\Delta\lambda_{\mathrm{PPS}}$ (see, e.g. paper [31]). It turned out that even for the unperturbed profile, where the average electron density $N_e^0 = 0.6N_c = 1.0 \times 10^{21}$ cm^{-3} is higher than for the perturbed profile, we have $\Delta\lambda_{\mathrm{EOS}} \sim \Delta\lambda_{\mathrm{PPS}} \sim 0.04$ mA, that is much smaller than 1.2 mA. Besides, the EOS is to the blue side, while the plasma polarization shift is to the red side, so that they essentially cancel each other out. (For completeness we note that in other experimental conditions the EOS could be to the red side—depending on the ratio of the oscillatory and thermal components of the electron velocity—and it could play a much more significant role for other spectral lines and/or conditions [3].)

Finally we note that possible dielectronic satellites do not seem to be relevant for the interpretation of the experimental profile. Indeed, in accordance with the calculations presented in paper [18], the most intense dielectronic satellite would be located at +9.3 mA at the red side. However, the experimental profile does not show any peak at this location. All other dielectronic satellites would be much further in the red wing, where the experimental profile merges with the noise. Also, we estimated a possible modification of laser satellites due to the mixing of singlet and triplet terms of the $n = 3$ multiplet—following section 5.1.5 of book [3]—and found it to be negligible.

In summary, in paper [16] we presented a complex analysis of the experimental Al Heβ emission from aluminum plasma created by one ps-laser beam and then subjected to another delayed ps-laser beam. The analysis was based on the advanced simulations coupling the code based on the Floquet–Liouville formalism with the particle-in-cell (PIC) kinetic code that provides a spatial distribution of the QEF in the plasma. We demonstrated that because the QEF in plasma was significantly different from the laser field in vacuum due to Raman and Brillouin backscattering, *the allowance for the spatial distribution was crucial for the adequate interpretation of the observed spectra—especially, for the identification of laser satellites.*

To conclude, by using the ultra-high resolution K-shell spectroscopy and advanced simulations, *we created a bridge between the physics of atomic processes in plasmas and the physics of laser–plasma interactions.* By doing so, the authors of paper [16] linked together the two corresponding research communities.

References

[1] Gavrilenko V P and Oks E 1987 *Sov. Phys. Plasma Phys.* **13** 22

[2] Oks E, Böddeker St and Kunze H-J 1991 *Phys. Rev.* A **44** 8338

[3] Oks E 1995 *Plasma Spectroscopy: The Influence of Microwave and Laser Fields. Springer Series on Atoms and Plasmas* vol 9 (New York: Springer) sections 4.2, 7.2–7.4

[4] Letokhov V S and Chebotaev V P 1977 *Nonlinear Laser Spectroscopy* (New York: Springer)

[5] Demtröder W 2003 *Laser Spectroscopy. Basic Concepts and Instrumentation* (New York: Springer)

[6] Zhuzhunashvili A I and Oks E 1977 *Sov. Phys. JETP* **46** 1122

[7] Oks E and Rantsev-Kartinov V A 1980 *Sov. Phys. JETP* **52** 50

[8] Gavrilenko V P and Oks E 1981 *Sov. Phys. JETP* **53** 1122

[9] Renner O, Dalimier E, Oks E, Krasniqi F, Dufour E, Schott R and Förster E 2006 *J. Quant. Spectrosc. Radiat. Transf.* **99** 439

[10] Jian L, Shali X, Qingguo Y, Lifeng L and Yufen W 2013 *J. Quant. Spectrosc. Radiat. Transf.* **116** 41

[11] Nguen H, Koenig M, Benredjem D, Caby M and Couland G 1986 *Phys. Rev.* A **33** 1279

[12] Salzmann D, Renner O, Sondhaus P, Förster E and Djaoui A 1997 *J. Quant. Spectrosc. Radiat. Transf.* **58** 851

[13] Baranger M and Mozer B 1961 *Phys. Rev.* **123** 25

[14] Cooper W S and Ringler H 1969 *Phys. Rev.* **179** 226

[15] Oks E and Gavrilenko V P 1983 *Sov. Phys. Tech. Lett.* **9** 111

[16] Sauvan P, Dalimier E, Oks E, Renner O, Weber S and Riconda C 2009 *J. Phys. B: At. Mol. Opt. Phys.* **42** 195001

[17] Ziener C *et al* 2002 *Phys. Rev.* E **65** 066411

[18] Renner O *et al* 2009 *High Energy Density Phys.* **5** 139

[19] Breil J and Maire P-H 2007 *J. Comput. Phys.* **224** 785

[20] Renner O, Uschmann I and Förster E 2004 *Laser Part. Beams* **22** 25

[21] Sauvan P and Dalimier E 2009 *Phys. Rev.* E **79** 036405

[22] Kolb A C and Griem H 1958 *Phys. Rev.* **111** 514

[23] Calisti A, Khelfaoui F, Stamm R, Talin B and Lee R W 1990 *Phys. Rev.* A **42** 5433

[24] Riconda C, Weber S, Tikhonchuk V, Adam J-C and Heron A 2006 *Phys. Plasmas* **13** 083103

[25] Weber S, Riconda C and Tikhonchuk V 2005 *Phys. Rev. Lett.* **44** 055005

[26] Gaysinsky I M and Oks E 1989 *J. Quant. Spectrosc. Radiat. Transf.* **41** 235

[27] Sauvan P, Dalimier E, Oks E, Renner O, Weber S and Riconda C 2010 *Int. Rev. At. Mol. Phys.* **1** 123

[28] Weber S, Riconda C and Tikhonchuk V 2005 *Phys. Plasmas* **12** 043101

[29] Oks E 1984 *Sov. Phys. Doklady* **29** 224

[30] Gaisinsky I M and Oks E 1986 *Correlations and Relativistic Effects in Atoms and Ions* (Moscow: USSR Acad. Sci. Research Council on Spectroscopy) p 106

[31] Koenig M, Malnoult P and Nguyen H 1988 *Phys. Rev.* A **38** 2089

Chapter 4

Spectroscopic diagnostics of relativistic laser–plasma interactions

4.1 L-dip phenomenon helps revealing parametric decay instabilities in laboratory and possibly astrophysical plasmas

Adapted from [1] [© 2020 Optical Society of America]. Users may use, reuse, and build upon the article, or use the article for text or data mining, so long as such uses are for non-commercial purposes and appropriate attribution is maintained. All other rights are reserved.

In paper [1] the authors analyzed the experimental spectra of Si XIV and Al XIII lines generated via the interaction of a super-intense (relativistic) laser radiation (of the intensity ~10^{21} W cm^{-2}) with thin Si foils. They discovered the ion acoustic turbulence in this laser-produced plasma. This was the first experimental discovery of the ion acoustic turbulence in laser-produced plasmas by means of the x-ray spectroscopy. The authors of paper [1] proved that the ion acoustic turbulence developed at the surface of the relativistic critical density as a result of the parametric decay instability (PDI).

PDI is a nonlinear process. In this process an electromagnetic wave decays into a Langmuir wave and an ion acoustic wave. Both waves originated from such decay represent two types of the electrostatic plasma turbulence. Electrostatic turbulence often takes place in various kinds of laboratory and astrophysical plasmas [2, 3]. It is represented by oscillatory electric fields. These fields relate to collective degrees of freedom in plasmas—in distinction to the electron and ion microfields that relate to individual degrees of freedom of charged particles.

In the absence of a magnetic field, low-frequency electrostatic plasma turbulence (LET), is represented only by ion-acoustic waves (also known as ionic sound waves). They have a broad spectrum at frequencies below or of the order of the ion plasma frequency $\omega_{\mathrm{pi}} = (4\pi e^2 N_i Z^2/m_i)^{1/2} = 1.32 \times 10^3\ Z(N_i m_p/m_i)^{1/2}$. Here N_i is the ion density, Z is the charge state; m_p and m_i are the proton and ion masses, respectively.

As for the high-frequency plasma turbulence, i.e. Langmuir waves, it develops at the electron plasma frequency $\omega_{pe} = (4\pi e^2 N_e/m_e)^{1/2} = 5.64 \times 10^4 \, N_e^{1/2}$, where N_e is the electron density. CGS units are used in the above formulas.

Langmuir waves were diagnosed by analyzing the shapes of neutral hydrogen spectral lines in many reliable experiments at various plasma sources and electron densities $N_e \sim 10^{14}$–10^{18} cm^{-3}. Specifically, Langmuir waves were diagnosed in a theta-pinch [4], Z-pinches [5, 6], and a gas-liner pinch [7] by using the phenomenon of the Langmuir-wave-caused dips (L-dips) in spectral line profiles. The majority of experimental studies, together with theoretical studies of L-dips are summarized in book [8]. For the latest results see chapter 3 of the present book.

As for the LET, it significantly modifies transport phenomena in various plasmas [9, 10]. For example, the LET in plasmas frequently leads to anomalous resistivity. In its turn, anomalous resistivity strongly changes the behavior of such plasmas.

In plasmas of low electron densities ($N_e \sim 10^{14}$–10^{16} cm^{-3}), different kinds of LET (e.g. ion-acoustic waves, Bernstein modes) were detected by the so-called 'anomalous' broadening of neutral hydrogen spectral lines. Various types of discharges were employed in these experiments [11–19]. The LET was also detected spectroscopically in solar flares. This was achieved by analyzing the shapes of the observed spectral lines of neutral hydrogen [20]. The same qualitative result was obtained also in experiment [21] devoted to the laboratory modeling of mechanisms of solar flares.

In dense plasmas the electrostatic waves/turbulence, such as, e.g. Langmuir waves and ion acoustic waves, can also develop. Specifically, in the research area of laser–plasma interactions, the electrostatic waves/turbulence have been studied theoretically by many authors: these studies included the relativistic laser interactions with solid targets (see, e.g. [22, 23] and references therein).

As for the first experimental discovery of the Langmuir waves in high-density laser-produced plasmas, this was achieved in work [24]. The authors of paper [24] utilized the phenomenon of L-dips in spectral line profiles of Al XIII. In distinction, ion acoustic waves (or any other kind of a LET) in high-density laser-produced plasmas have not been discovered experimentally until work [1].

It is important to emphasize that the authors of paper [24] showed that the ratio of the energy density of the turbulent electric fields $E^2/(8\pi)$ to the thermal energy density of the plasma $N_e T$ (where T is the plasma temperature) is of the same order of magnitude as the corresponding ratio in different astrophysical objects: quasars, pulsars, and Seyfert galaxies. Consequently, experiments like this in laboratory conditions can serve as a tool to model the PDI in the above astrophysical objects.

These very different astrophysical objects have one common feature: in their atmosphere, the conditions are satisfied for the PDI, causing the anomalous absorption of the incident electromagnetic radiation—see, e.g. [25].

This possibility, i.e. utilizing the x-ray spectroscopy in laboratory laser–plasma interaction experiments as a tool to model the PDI in the above astrophysical objects, is based on the same principle (the similarity of dimensionless parameters) that was employed in various modeling experiments at plasmas not produced by lasers. For example, this principle of modeling was used for decades for the successful modeling of mechanisms of solar flares by laboratory experiments—see, e.g. review [26] and references therein. While the dimensional parameters in the two objects differed by many orders of magnitude, the relevant dimensionless parameters were of the same order of magnitude—see, e.g. paper [27].

The experiments described in paper [1] were performed at Vulcan Petawatt facility located at the Rutherford Appleton Laboratory [28, 29]. Using optical parametric, chirped pulse amplification (OPCPA) technology, it produced a laser beam with a central wavelength of 1054 nm and a pulse full-width-half-maximum (FWHM) duration, which could be varied from 500 up to 1500 fs. Due to the OPCPA approach, it was possible to achieve an amplified spontaneous emission-to-peak-intensity contrast ratio exceeding $1:10^9$ several nanoseconds before the peak of the laser pulse [30]. The laser pulse was focused with an f/3 off-axis parabola. At the best focus approximately 30% of the energy (up to 290 J on the target in the experiments reported in paper [1]) was contained within a 7 μm (FWHM) diameter spot providing a maximum intensity of 1.4×10^{21} W cm^{-2}. The horizontally polarized laser beam was incident on target at 45° from the target surface normal, as shown schematically in figure 4.1(a).

With the help of a focusing spectrometer with spatial resolution (FSSR) [31–33], high-resolution spectroscopy measurements were performed [1]. The instrument was equipped with a spherically bent mica crystal with a lattice spacing $2d \sim 19.94$ Å and a radius of curvature of $R = 100$ mm. The crystal was aligned to operate at $m = 3$ order of reflection to record K-shell emission Rydberg H-like spectral lines of multicharged Si XIV ions in 4.85–5.35 Å wavelength range. The FSSR spectral resolving power was approximately 3000. The spectrometer viewed the laser-irradiated rear surface of the target at an angle of 5° to the target surface normal (figure 4.1(a)). Target-to-crystal distance of 367 mm led to a demagnification factor of ~4.7 for the diffracting system.

A pair of 0.5 T neodymium–iron–boron permanent magnets, which formed a slit of 10 mm wide, was placed in front of the crystal. The purpose was to reduce the level of noise caused by the background fogging and crystal fluorescence. Besides, for registering spectra the authors of paper [1] utilized an image plate that was not sensitive to the electromagnetic noise. A great advantage of the measurements presented in paper [1] was the employment of a high luminosity FSSR (see paper [31] for the details). This made it possible to dramatically enhance the signal/noise ratio. Indeed, the entire intensity of the x-ray spectra emitted by plasma in the direction perpendicular to the spectra dispersion, was focused in the line of some pxl thickness. Due to this, the signal/noise ratio got significantly increased. All the above mentioned arrangements allowed measuring spectra in a single laser shot with a high spectral resolution and a high signal/noise ratio.

Spectra were recorded on Fujifilm TR image plate, which was protected against exposure to visible light using two layers of 1 μm thick polypropylene filter coated with 0.2 μm Al. In addition to these, a 1 μm thick polypropylene filter was placed at the magnet slit. The spectra were measured from the rear side of the foil: the most intense contribution to the spectral lines originated from the densest part of the plasma, i.e. from the region near the relativistic critical density. It should be emphasized that any spectra of plasma, produced by amplified spontaneous

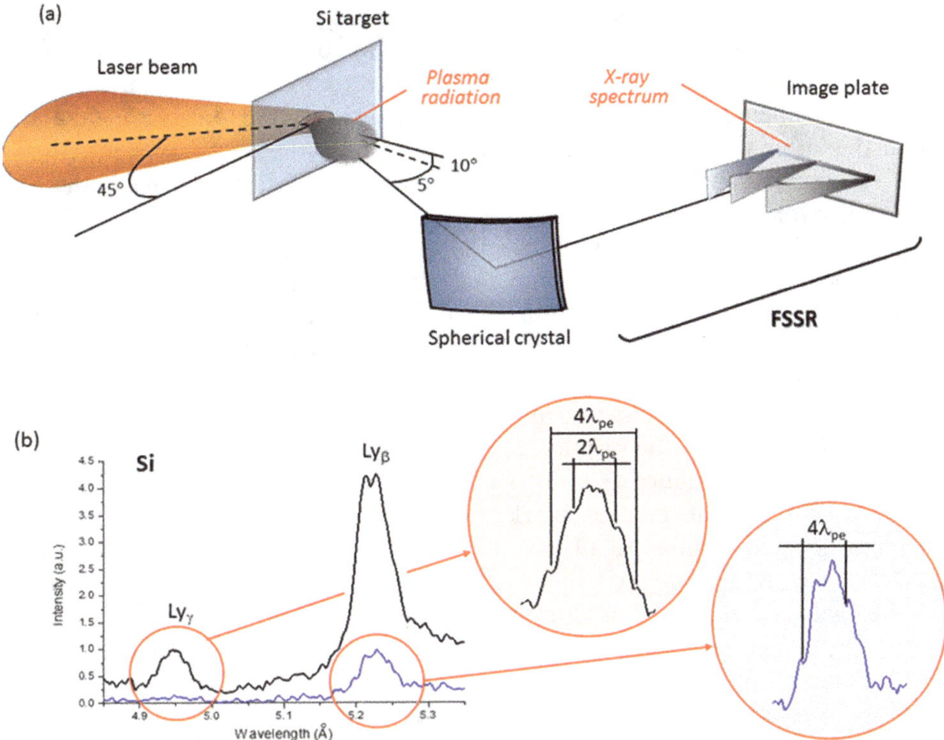

Figure 4.1. Schematic of experimental setup and typical x-ray spectra in the range of 0.485–0.535 nm [1]. Experimental setup (a) and profiles (b) of Si XIV spectral lines, obtained in a single laser shot with initial laser intensity at the surface of the target estimated as 1.01×10^{21} W cm^{-2} (black trace) and 0.24×10^{21} W cm^{-2} (blue trace). In the insets, positions of the dips/depressions in the profiles are marked by vertical lines separated either by $2\lambda_{pe}$ or $4\lambda_{pe}$, where $\lambda_{pe} = [\lambda_0^2/(2\pi c)]\omega_{pe}$ (λ_0 is the unperturbed wavelength of the corresponding line). Reproduced from [1] with permission from Optical Society of America.

emission at the electron density lower than the non-relativistic critical density ($<10^{21}$ W cm^{-2}) have not been observed by both front and rear side spectrometers —see for details the experimental results from paper [34] obtained at the same experimental conditions. It should be underlined that the experimental spectra of Si XIV, presented in paper [1], were observed from the rear side of target, what suggests that no influence of the amplified spontaneous plasma emission should be important.

Figure 4.1(b) exhibits the experimental spectra of Si XIV Ly$_\beta$ and Ly$_\gamma$ lines in shots 72 (2 μm Si foil) and 89 (1 μm Si foil). They were obtained in a single laser shot with duration of 0.6 ps and laser intensities at the surface of the target theoretically estimated as 1.01×10^{21} W cm^{-2} and 0.24×10^{21} W cm^{-2}, respectively. The positions of the dips/depressions in the profiles are marked in insets by vertical lines separated either by $2\lambda_{pe}$ or $4\lambda_{pe}$, where $\lambda_{pe} = [\lambda_0^2/(2\pi c)]\omega_{pe}$ (λ_0 is the unperturbed wavelength of the corresponding line).

Before proceeding to the analysis of the experimental spectra, we need to add some details to the theory of the L-dips briefly presented in section 3.1. The L-dips originates from a resonance between many quasienergy harmonics of the combined system 'radiator + oscillatory field' caused simultaneously by all harmonics of the total electric field $\mathbf{E}(t) = \mathbf{F} + \mathbf{E}_0 \cos \omega_{pe} t$, where vectors $\mathbf{F}$ and $\mathbf{E}_0$ are not collinear. The quasistatic field $\mathbf{F}$ represents the low-frequency part of the ion microfield and the LET, if the latter was developed in the plasma.

The resonance condition is $\omega_F = s\omega_{pe}(N_e)$, s = 1, 2, ..., where ω_F is the Stark splitting of hydrogenic energy levels, caused by a quasistatic field $\mathbf{F}$ in a plasma: $\omega_F = 3n\hbar F/(2Z_r m_e e)$. In the case where the quasistatic field $\mathbf{F}$ is dominated by the LET, for the Ly-lines the distance of an L-dip from the unperturbed wavelength λ_0 is given by $\Delta\lambda_{\mathrm{dip}}(q, N_e) = -[\lambda_0^2/(2\pi c)]qs\omega_{pe}(N_e)$. Here $q = n_1 - n_2$ is the electric quantum number expressed via the parabolic quantum numbers n_1 and n_2: $q = 0, \pm1, \pm2, ...,$ $\pm(n-1)$. It labels Stark components of Ly-lines. For a pair of Stark components, corresponding to q and $-q$, there could be a pair of L-dips located symmetrically in the red and blue parts of the profile: $\Delta\lambda_{\mathrm{dip}}(N_e) = \pm[\lambda_0^2/(2\pi c)]qs\omega_{pe}(N_e)$ for the case of the one quantum resonance ($s = 1$). For the two quantum resonance ($s = 2$), in the profile of the same pair of Stark components, there could be another pair of L-dips located symmetrically in the red and blue parts of the profile: $\Delta\lambda_{\mathrm{dip}}(N_e) = \pm[\lambda_0^2/(2\pi c)]q\omega_{pe}(N_e)$.

If the quasi-static field $\mathbf{F}$ is dominated by the ion microfield, then the above formula for $\Delta\lambda_{\mathrm{dip}}$ would hold only for relatively low electron densities. For relatively high electron densities, the spatial non-uniformity of the ion microfield has to be taken into account. The corresponding result is represented in equation (3.5).

In any case (whether the field $\mathbf{F}$ is dominated by the ion microfield or by a LET), the separation of the two L-dips, corresponding to q and $-q$, is

$$\Delta\lambda_{\mathrm{dip}}(-|q|, N_e) - \Delta\lambda_{\mathrm{dip}}(|q|, N_e) = [\lambda_0^2/(\pi c)] \, |q| s\omega_{pe}(N_e), \qquad (4.1)$$

thus allowing us to measure N_e. It is worth pointing out that this passive spectroscopic method for measuring N_e is just as accurate as the active spectroscopic method (more complicated experimentally) using the Thompson scattering, as shown in the benchmark experiment [7]. The half-width of the L-dip (i.e. the separation between the dip and the nearest 'bump'), is controlled by the amplitude E_0 of the Langmuir wave [8]:

$$\delta\lambda_{1/2} \approx \left(\frac{3}{2}\right)^{1/2} \frac{\lambda_0^2 n^2 E_0}{8\pi m_e ec Z_r} \qquad (4.2)$$

Consequently, it is possible to deduce the amplitude E_0 of the Langmuir wave by measuring the experimental half-width of L-dips.

Before presenting the analysis of the experimental Si XIV spectral lines, the authors of paper [1] noted the following. The laser frequency ω corresponds to the wavelength λ of approximately 1054 nm. At lower, non-relativistic laser intensities, the critical electron density N_c is determined from the equation $\omega = \omega_{pe}$, where $\omega_{pe} = (4\pi e^2 N_e/m_e)^{1/2}$, would be 1.0×10^{21} cm^{-3}. However, at the laser intensities ~10^{21} W cm^{-2}, i.e. those corresponding to the experiment [24], due to relativistic effects, the 'relativistic' critical electron density N_{cr} becomes higher than N_c—see, e.g. papers [35–37]. For the linearly-polarized laser radiation, it becomes (according to paper [38])

$$N_{\mathrm{cr}} = \frac{(\pi a/4)m_e\omega^2}{4\pi e^2}, \quad a = \lambda(\mu m)\left[\frac{I(\mathrm{Wcm}^{-2})}{1.37 \times 10^{18}}\right]^{1/2} \tag{4.3}$$

Equation (4.3) is the quantitative result of the phenomenon called 'relativistically-induced transparency'. The latter is the ability of the ultra-intense laser radiation to penetrate into regions of the density higher than N_c. This has been previously demonstrated by PIC simulations [39]. In [1] the authors not only took into account the effect of the relativistically induced transparency (details of which can be found in paper [40]) on the resonance condition, but also confirmed that the L-dips originate from the region of the relativistic critical density (which is greater than the nonrelativistic critical density), as shown below.

Let us consider the experimental profile of Si XIV Ly-gamma, produced in an interaction with a single laser pulse with initial laser intensity at the surface of the target estimated as 1.01×10^{21} (black trace in figure 4.1(b)). The experimental profile shows a pair of two L-dips nearest to the line center ($q = \pm1$, $s = 1$) separated from each other by 28 mÅ. From this separation, the authors of paper [1] deduced the electron density of $N_e = 3.6 \times 10^{22}$ cm^{-3}. In addition to this, a pair of L-dips, separated from each other by 56 mÅ, were also observed, representing a super-position of two pairs of the L-dips: $q = \pm2$, $s = 1$ and $q = \pm1$, $s = 2$. From the separation of the L-dips within this pair, the authors of paper [1] deduced the same $N_e = 3.6 \times 10^{22}$ cm^{-3}. The fact that the analysis of the two different pairs of the L-dips resulted in the same electron density reinforced the interpretation of the experimental dips as the actual L-dips—caused by the resonant interaction of the Langmuir waves, developed at the surface of the relativistic critical density, with the quasistatic electric field.

In both cases, L-dips separated by 28 mÅ and for the L-dips separated by 56 mÅ, the mid-point between the two dips in the pair practically coincides with the unperturbed wavelength λ_0. This was a strong indication that the quasistatic field **F** was dominated by the LET. If the LET was absent, then according to equation (3.5), the mid-point of the pair of the L-dips separated by 28 mÅ should have been shifted by 5.8 mÅ to the red with respect to λ_0 and the mid-point of the pair of the L-dips separated by 56 mÅ would have been similarly shifted by 10.7 mÅ to the red with respect to λ_0, these shifts being due to the spatial non-uniformity of the ion microfield reflected by the second term in equation (3.5).

Another strong indication of the presence of the LET observed from the above result comes from the analysis of the broadening of this spectral line that was performed using code FLYCHK [41]. This code, which does not take into account the Stark broadening by the LET (or the presence of L-dips), yielded $N_e = 0.9 \times 10^{23}$ cm^{-3}, i.e. almost three times higher than the actual $N_e = 3.6 \times 10^{22}$ cm^{-3} (the best fit to the experimental profile by code FLYCHK is shown in figure 4.2(a)). This is yet another strong indication of an additional Stark broadening by the LET (not accounted for by FLYCHK).

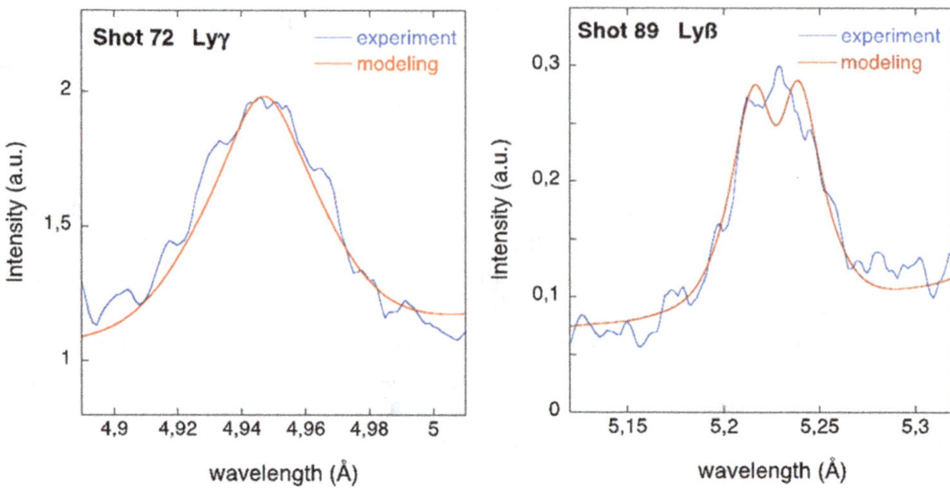

Figure 4.2. Experimental spectra and their comparisons with FLYCHK modeling [1]. (a) Comparison of the experimental profile of Si XIV Ly$_\gamma$ line in shot with initial laser intensity at the surface of the target estimated as 1.01×10^{21} W cm^{-2} with a simulation performed using a variation of the code FLYCHK for calculating the Stark broadening, then adding both Doppler and instrument broadening, and, if necessary, opacity. The L-dips phenomenon and the spectral line broadening by LET were not included in the FLYCHK. The plasma parameters for the best fit are $N_e = 0.9 \times 10^{23}$ cm^{-3} and $T_e = 500$ eV. (b) The same as in (a) but for Si XIV Ly$_\beta$ line produced in a single laser shot with initial laser intensity at the surface of the target estimated as 0.24×10^{21} W cm^{-2}; $N_e = 3 \times 10^{23}$ cm^{-3} and $T_e = 500$ eV. Reproduced from [1] with permission from Optical Society of America.

At the initial laser intensity at the surface of the target estimated as 1.01×10^{21} W cm^{-2}, the relativistic critical density would be $N_{cr} = 2.3 \times 10^{22}$ cm^{-3} according to equation (4.3). However, due to several physical effects, the actual intensity of the transverse electromagnetic wave in the plasma can be significantly higher than the intensity of the incident laser radiation at the surface of the target. One of these effects is the self-focusing of the laser beam in plasmas—details of the process of the laser propagation in the plasma corona at the overcritical density can be found, for example, in papers [40, 42] and in reviews [22, 43]. The other effects are Raman and Brillouin backscattering, For example, in paper [44] this was proven by a spectroscopic analysis of the experimental x-ray line profiles and by PIC simulations. In the experiment under consideration, in shot 72, for the relativistic critical density to be approximately equal to the density $N_e = 3.6 \times 10^{22}$ cm^{-3} deduced by the spectroscopic analysis, it would require the enhancement of the initially estimated intensity of the transverse electromagnetic wave at the surface of the target due to the above physical effects just by a factor of two.

The authors of paper [1] also evaluated (independently of the spectroscopic analysis) the enhancement of laser intensity due to the self-focusing from the well-known model—see, e.g. chapter 4 of book [45]. Taking into account our plasma parameters one can get the enhancement of the laser intensity by the factor of two, which was consistent with the results of our spectroscopic analysis.

Previous studies have shown (see, e.g. [22, 23]) that at high laser intensities, the most probable and the best studied mechanism for developing Langmuir waves at the surface of the relativistic critical density is a parametric decay, which is a nonlinear process where the pump wave (t_1) excites both the Langmuir wave (l) and an ion-acoustic wave (s): $t_1 \rightarrow l + s$. In shot 72, the authors of paper [1] experimentally discovered LET, which developed simultaneously with the Langmuir waves. Therefore, first, the LET should have been also developed at the surface of the relativistic critical density and thus should have been most probably the ion acoustic turbulence. Second, the self-focusing, as well as Raman and Brillouin backscattering most probably enhanced the intensity of the transverse electromagnetic wave in the plasma by a factor of two. To further prove these conclusions, the authors of paper [1] performed also PIC simulations at the electromagnetic wave intensity enhanced by a factor of two. These simulations, presented below, confirmed the development of both Langmuir and ion acoustic waves.

It should be mentioned that L-dips were not observed in the experimental profiles of Si XIV Ly$_\alpha$ and Ly$_\beta$ lines in shot 72. It seems that a process of self-absorption of radiation prevented L-dips from being visible in these lines: the self-absorption 'washes out' fine features of the experimental profiles.

The analysis of the experimental profile of Si XIV Ly$_\beta$ in shot 89 with the initial laser intensity at the surface of the target estimated as 2.4×10^{20} W cm^{-2} (blue trace in figure 4.1(b)), shows a situation similar to Si XIV Ly$_\gamma$ in shot 72. There is a pair of the L-dips separated from each other by 43 mA. The electron density deduced from the separation within the pair of these L-dips, is $N_e = 1.74 \times 10^{22}$ cm^{-3} (assuming $|q|s = 2$). The location of the would-be L-dips, corresponding to $|q|s = 1$, is too close to the central, most intense part of the experimental line profile: they are not observed either due to a self-absorption in the most intense part of the profile, or because the relatively small values of the field $\mathbf{F}$, corresponding to the central part of the profile, are not quasistatic (as is well-known from review [46]). So, the would-be L-dips, corresponding to $|q|s = 1$ could not form for either or both reasons.

The mid-point between the two dips in the pair practically coincides with the unperturbed wavelength λ_0. This was again a strong indication that the quasistatic field $\mathbf{F}$ was dominated by the LET. If the LET were absent, then according to equation (3.5), the mid-point of this pair of the L-dips should have been shifted by 6.1 mÅ to the red with respect to λ_0.

Another strong indication of the presence of the LET in shot 89 comes from the analysis of the broadening of this spectral line following from the modeling, which was performed using code FLYCHK [41]. An electron density of $N_e = 3 \times 10^{23}$ cm^{-3}, was obtained, i.e. 17 times higher than the experimentally verified $N_e = 1.74 \times 10^{22}$ cm^{-3} (the best fit to the experimental profile by code FLYCHK is shown in figure 4.2(b)). This is yet another strong indication of an additional Stark broadening by the LET (not accounted for by FLYCHK).

At the initial laser intensity at the surface of the target estimated as 2.4×10^{20} W cm^{-2}, the relativistic critical density would be $N_{cr} = 1.1 \times 10^{22}$ cm^{-3} according to equation (4.3). However, again due to the self-focusing, as well as Raman and

Brillouin backscattering, the actual intensity of the transverse electromagnetic wave in the plasma can be significantly higher. In the experiment under consideration, in shot 89, for the relativistic critical density to be approximately equal to the density $N_e = 1.74 \times 10^{22}$ cm^{-3} deduced by the spectroscopic analysis, again it would require the enhancement of the intensity of the transverse electromagnetic wave due to the above physical effects just by a factor of two. By the same reasoning as presented above with regard to shot 72, the experimentally discovered LET in shot 89, which developed simultaneously with the Langmuir waves, should also have been developed at the relativistic critical density surface and thus is most likely to be ion acoustic turbulence.

In shot 89, L-dips were not observed in the experimental profiles of Si XIV Ly$_\alpha$ and Ly$_\gamma$ lines. This is because the Ly$_\alpha$ line seemed to experience a significant self-absorption, which prevented L-dips from being visible. As for the Ly$_\gamma$ line in shot 89, at the possible locations of the L-dips, the experimental profile already merged into the noise.

For a further quantitative analysis, the authors of paper [1] calculated the corresponding theoretical profiles, where the total quasistatic field is $\mathbf{F} = \mathbf{F}_t + \mathbf{F}_i$. For calculating the distributions of the total quasistatic field $\mathbf{F} = \mathbf{F}_t + \mathbf{F}_i$, where $\mathbf{F}_t$ is the field of a LET and $\mathbf{F}_i$ is the quasistatic part of the ion microfield, they employed the results from references [48, 49]: the distribution of the total quasistatic field is a convolution of the Rayleigh-type distribution of F_t [47] with the APEX distribution of F_i [49]. The broadening by the electron microfield, as well as the Doppler and instrumental broadenings were also taken into account. For calculating the spectra in the regions of L-dips the authors of paper [1] used the analytical solution [8, 50] for the wave functions of the quasienergy states caused simultaneously by all harmonics of the total electric field $\mathbf{E}(t) = \mathbf{F} + \mathbf{E}_0 \cos \omega_{pe}t$, where vectors $\mathbf{F}$ and $\mathbf{E}_0$ are not collinear. Further details can be found in paper [50] and in book [8], section 4.2. Below is the outcome.

From the experimental profile of Si XIV Ly-gamma in shot 72, it follows that the root-mean-square value of F_t was $F_{t,\mathrm{rms}} = 2.1$ GV cm^{-1}. For comparison, the characteristic ion microfield $F_{i,\mathrm{typ}} = 2.603\ eZ^{1/3} N_e^{2/3}$ was 1.0 GV cm^{-1}. From the half width of the experimental L-dips, by using equation (4.2), the authors of paper [1] found the amplitude of the Langmuir wave to be $E_0 = 0.6$ GV cm^{-1}. The resonant value of the quasistatic field F_{res}, determined by the condition of the resonance between the separation of the Stark sublevels and the plasma frequency $3n\hbar F_{\mathrm{res}}/(2Z_r m_e e) = \omega_{pe}$, was 3.1 GV cm^{-1}, so that the validity condition for the existence of L-dips $E_0 < F_{\mathrm{res}}$ [8] was satisfied.

From the experimental profile of Si XIV Ly-beta in shot 89, it follows that the root-mean-square value of F_t was $F_{t,\mathrm{rms}} = 3.9$ GV cm^{-1}. For comparison, the characteristic ion microfield was $F_{i,\mathrm{typ}} = 0.6$ GV cm^{-1}. From the halfwidth of the experimental L-dips, by using equation (4.2), the authors of paper [1] found the amplitude of the Langmuir wave to be $E_0 = 1.0$ GV cm^{-1}. The resonant value of the quasistatic field was $F_{\mathrm{res}} = 5.8$ GV cm^{-1}, so that the validity condition for the existence of L-dips $E_0 < F_{\mathrm{res}}$ was satisfied.

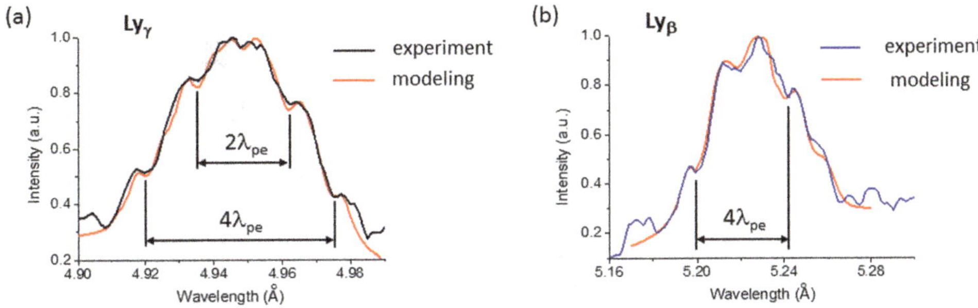

Figure 4.3. Experimental spectra and their comparisons with a different Stark broadening code including L-dips and the spectral line broadening by LET [1]. (a) Experimental spectra of Si XIV Ly_γ line in a single laser shot with initial laser intensity at the surface of the target estimated as 1.01×10^{21} W cm^{-2} initial laser intensity at the surface of the target. The positions of the dips/depressions in the profiles are marked by vertical lines separated either by $2\lambda_{pe}$ or $4\lambda_{pe}$, where $\lambda_{pe} = [\lambda_0^2/(2\pi c)]\omega_{pe}$ (λ_0 is the unperturbed wavelength of the corresponding line). Also shown is a theoretical profile at $N_e = 3.6 \times 10^{22}$ cm^{-3} allowing, in particular, for a low-frequency electrostatic turbulence (see the text for details). (b) Same but for Si XIV Ly_β line in a single laser shot with initial laser intensity at the surface of the target estimated as 0.24×10^{21} W cm^{-2} initial laser intensity at the surface of the target. The theoretical profile is shown at $N_e = 1.74 \times 10^{22}$ cm^{-3}. Reproduced from [1] with permission from Optical Society of America.

The comparison of the theoretical profiles, allowing, in particular, for a LET and L-dips, with the corresponding experimental profiles is shown in figure 4.3.

The good agreement between experimental spectra and simulations reinforces the discovery of the simultaneous production of LET with the Langmuir waves. It should be noted that the electron densities involved turned out to be much lower than the densities deduced using FLYCHK simulations, which ignored the LET and the L-dips.

The authors of paper [1] also performed experiments where the spectrometer viewed the laser-irradiated *front* surface of the target. As an example, figure 4.4 shows the experimental spectrum of Al XIII Ly_β line in shots 82 (4 μm Al foil coated by 0.45 μm CH), which was obtained in a single laser shot with duration of 0.9 ps and the laser intensity at the surface of the target theoretically estimated as 6.7×10^{20} W cm^{-2}.

This spectrum exhibits two pairs of L-dips: one pair—at ±16.8 mA from the line center, another pair—at ±33.6 mA from the line center. The L-dips are very pronounced: for example, for the L-dips at ±16.8 mA from the line center, the intensity of the bump nearest to the center of the dip exceeds the intensity at the center of the dip by 15% for the L-dip at +16.8 mA and by 11% for the L-dip at −16.8 mA. Just by itself, such a strong modulation of the line profile clearly indicates that it is not a noise. More importantly—and this argument concerns all of the above spectra—the fact that the observed dips are located *symmetrically* (rather than randomly) with respect to the line center (located at 6.06 A) indicates that this is not a noise. This is especially clear in cases where two pairs of symmetrically-located L-dips were observed (such as in Al XIII Ly_β spectrum in shot 82 and in Si XIV Ly_γ spectrum in shot 72): not only there was the symmetry (rather than randomness) in terms of the dips location within each pair, but also the distance from the line center

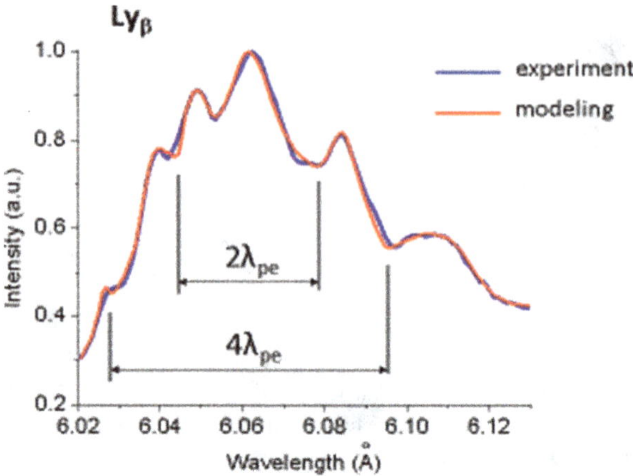

Figure 4.4. Experimental spectrum of Al XIII Ly$_\beta$ and its comparison with the Stark broadening code including L-dips and the spectral line broadening by LET [1]. The experimental spectrum was obtained in a single laser shot with initial laser intensity at the surface of the target estimated as 6.7×10^{20} W cm^{-2}. The positions of the L-dips in the profiles (the L-dips being very pronounced) are marked by vertical lines separated either by $2\lambda_{pe}$ or $4\lambda_{pe}$, where $\lambda_{pe} = [\lambda_0^2/(2\pi c)]\omega_{pe}$ (λ_0 is the unperturbed wavelength of the corresponding line). Also shown is a theoretical profile at $N_e = 2.35 \times 10^{22}$ cm^{-3} allowing, in particular, for L-dips and for a low-frequency electrostatic turbulence. This value of N_e is much lower than N_e obtained from fitting the same experimental spectrum by code FLYCHK that did not include the L-dips phenomenon and the spectral line broadening by LET, and therefore had significant discrepancies with the experimental profile at the locations of the L-dips. Reproduced from [1] with permission from Optical Society of America.

for one pair was consistent (differed exactly by the factor of 2, as theoretically expected) with the distance from the line center for the second pair.

Finally, it is worth noting that the characteristic lifetime of the upper state n of the radiating ion (i.e. of the state, from which the observed spectral line originates), being controlled by the electron dynamical broadening, is much smaller than the period of the LET. Therefore, each radiating ion 'transmits' to the spectrometer a snapshot of the instantaneous electric field of the LET at the location of the radiating atom. This is the essence and the justification of the quasistatic description of the LET. Different radiating ions are subjected to different values of the LET fieldstrength, so that the averaging/integration over the ensemble of the radiating ions conveys the distribution of the electric field of the LET. Since the radiating ions are constantly excited to the upper state n (followed by the decay) at various times during the laser pulse and at various locations, there are always radiating ions that 'witness' the LET resulting from the development of the parametric decay instability.

Also it should be emphasized that measurements of plasma parameters by intensity ratios of various lines of H and He-like ions of Si, processed for the same experimental conditions by Los Alamos code Atomic [34], clearly demonstrated that the plasma emission originated from the region of the electron density of about 10^{22} W cm^{-2}. Those results are in a good agreement with the results of paper

[1], the latter including the influence of the parametric instabilities on the x-ray spectra.

The authors of paper [1] supported the results of their spectroscopic analysis by particle-in-cell (PIC) simulations. The simulations were performed using the modified 1D PIC code LPIC, in which a laser pulse of duration $t_L = 0.6$ ps and intensity $I_L = 2.7 \times 10^{21}$ W cm^{-2} interacted with a Si [51]. Laser pulse propagates along the x-axis and interacts with Si^{+14} inhomogeneous plasma layer with a linear density ramp over the length $L = 6$ μm and a target thickness of 2 μm, with a constant ion density of $N_i = 6 \times 10^{22}$ cm^{-3}. Angle of incidence was set up 45° at P—polarization. The results of the simulations are shown in figure 4.5.

At the peak of the laser pulse intensity ($t = 320$ fs) it is seen that the scale of laser field decay is close to the scale of the plasma inhomogeneity. The longitudinal field of the Langmuir wave E$_l$ appeared in the vicinity of the point where the density is about 1/4 of the relativistic critical density. The Langmuir wave exists up to a point slightly above the relativistic critical density $N_{cr} = 3.6 \times 10^{22}$ cm^{-3}. In figure 4.5(b), the region near the relativistic critical density is $8\lambda < x < 8.25\lambda$: the modulation of ion density in this region shown by the green line is the manifestation of the ion acoustic wave. Similar processes of such parametric decays ($t \rightarrow l + l'$, s) were studied in the past (see, e.g. book [52]), but for significantly lower laser intensities.

In summary, in paper [1] the authors experimentally discovered LET, which developed simultaneously with the Langmuir waves. Both types of the waves have been developed at the surface of the relativistic critical density and thus the LET most probably have been the ion acoustic turbulence. These conclusions have been also supported by PIC simulations. The authors of paper [1] determined both the amplitude of the Langmuir waves and the root-mean-square field of the ion acoustic waves.

The ratio of the energy density of the turbulent electric fields $E^2/(8\pi)$ to the thermal energy density of the plasma N_eT in the experiment [1] was $\sim 10^{-1}$. In atmospheres of quasars, pulsars, and Seyfert galaxies, the electron density is at least by 12 orders of magnitude lower than in the above experiments, but the order of magnitude of the ratio $E^2/(8\pi N_eT)$ is similar to the order of magnitude of this ratio in the above experiments. Since the conditions for developing the PDI are satisfied for atmospheres of these astrophysical objects [25], the above type of laboratory experiments could serve as a tool for modeling the PDI in these astrophysical objects. The principle of modeling astrophysical objects in laboratory experiments based on the similarity of dimensionless controlling parameters was already successfully used, e.g. in laboratory modeling of mechanisms of solar flares (see, e.g. review [26]).

The experimental spectra presented in paper [1] were integrated over the pulse duration. Time-resolved x-ray spectroscopy had been already performed for similar experimental situations—see, e.g. paper [41]. In that paper it was shown from the Li-like emission history for Al thin foils that during 700 fs the *electron density* of the plasma, deduced from the broadening of the lines, was very high (solid-state density) and *steady*. It was attributed to the hottest plasma produced during the highest laser intensity. After 700 fs there was no significant signal that would come from colder

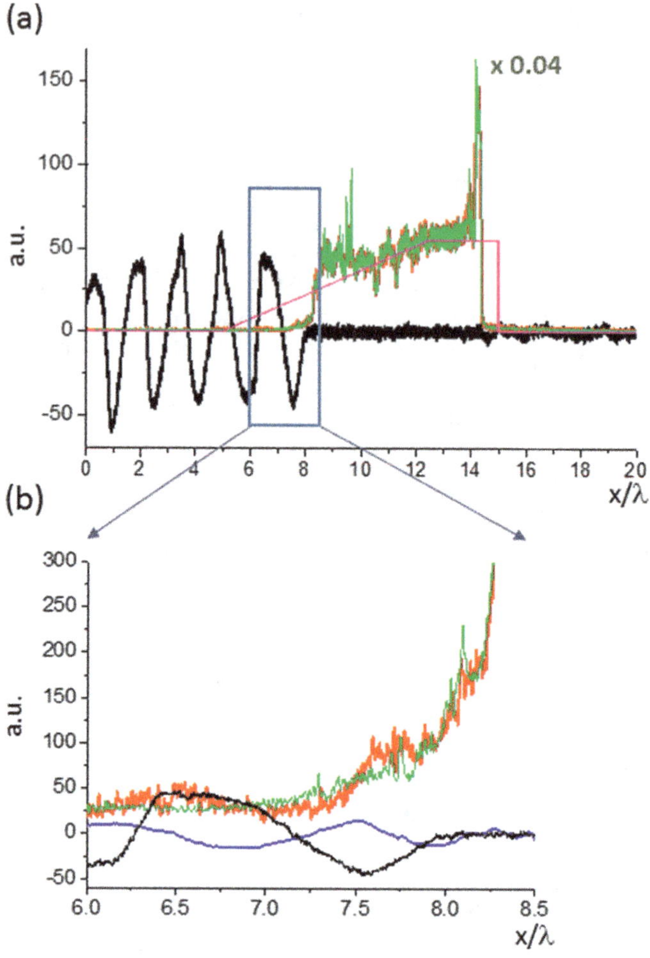

Figure 4.5. PIC simulations of the interaction of a 600 fs pulse of the intense (2.7×10^{21} W cm^{-2}) linearly-polarized laser radiation with the Si^{14} plasma layer [1]. The general view of the laser-target interaction at the instant $t = 320$ fs from the beginning of the interaction (a) and the zoom (b) on the area shown by the black rectangle in (a). The coordinate x is in units of the laser wavelength λ. The initial scaled density profile is presented by the violet line. The scaled electron and ion densities are presented by the red and green lines. They are given in the units of $0.04ZN_{\mathrm{cr}}$ in (a) and in the units of ZN_{cr} in (b); here $N_{\mathrm{cr}} = \gamma N_c$, where N_c is the critical plasma density and $\gamma = a_L$ is the relativistic factor. The calculated scaled transversal electric field $a_L = eE_L/(m_e\omega_L c)$ is shown by the black line and the longitudinal one (the Langmuir wave) $a_l = eE_l/(m_e\omega_L c)$—by the blue line. Reproduced from [1] with permission from Optical Society of America.

plasma. So, the authors of paper [1] interpreted their experimental results (in this regard) the same way as in paper [40].

The idea of employing these two different codes simultaneously, as was done by the authors of paper [1], should be applied to other spectroscopic experiments in laser-produced plasmas showing identifiable L-dips. The identification of L-dips in the experimental line profiles provides a very accurate determination of the electron density N_e independent of the broadening of the spectral lines. Then if a code like FLYCHK, ignoring the LET and the L-dips, yields a much higher value of N_e than the one deduced from the L-dips, this would be an indication of the development of the LET in those other experiments.

Thus, the results presented in paper [1] should be important from both theoretical and practical points of view. The experimental discovery of the ion acoustic turbulence in laser-plasma experiments is an important new result in its own right, enabling the comparison of theoretical predictions concerning the development of the ion acoustic waves with the experimentally determined parameters of these waves. The results presented in paper [1] can be also used for important practical purposes. One example is modeling physical processes in atmospheres of quasars, pulsars, and Seyfert galaxies—in particular, the PDI, causing the anomalous absorption of the incident electromagnetic radiation. Another example is providing a better understanding of the physics of laser-produced plasmas, in particular, the transport phenomena (especially, the anomalous resistivity)—as they are strongly affected by the ion acoustic waves.

4.2 In-depth study of intra-Stark spectroscopy in the x-ray range in relativistic laser–plasma interactions

In paper [53] the authors advanced the study of the relativistic laser–plasma interactions from paper [1] by having the following new features. First, in the experiment presented in paper [53] a plasma mirror was used to significantly improve the laser contrast. As a result, there was a much smaller laser pre-plasma (compared to the experiment from paper [1]), allowing the main laser pulse to interact with a higher density plasma.

Second, this improvement enabled the authors of paper [53] to perform an in-depth spectroscopic study of the simultaneous production of the Langmuir waves and of the ion acoustic turbulence at the surface of the relativistic critical density. This was achieved by the spectroscopic analysis of the same spectral line (Si XIV Ly-beta) in three different shots of about the same laser intensity 2×10^{20} W cm^{-2}. This demonstrated reliable reproducibility of the L-dips at the same locations in the experimental profiles, as well as of the deduced parameters (fields) of the Langmuir waves and ion acoustic turbulence in all three shots.

Third, the study presented in paper [53] employed for the first time *the most rigorous condition of the dynamic resonance* (which makes possible the ISS phenomenon) compared to all previous studies in all kinds of plasmas, over a wide range of electron densities. By doing so the authors of paper [53] showed how different interplays between the Langmuir wave field with the field of the ion acoustic turbulence lead to distinct spectral line profiles, including the disappearance of the L-dips.

The experiments reported in paper [53] were performed at the same Vulcan Petawatt (PW) laser facility at the Rutherford Appleton Laboratory, as the experiments presented in paper [24]. Vulcan PW generates a beam using optical parametric, chirped pulse amplification (OPCPA) technology at central wavelength of

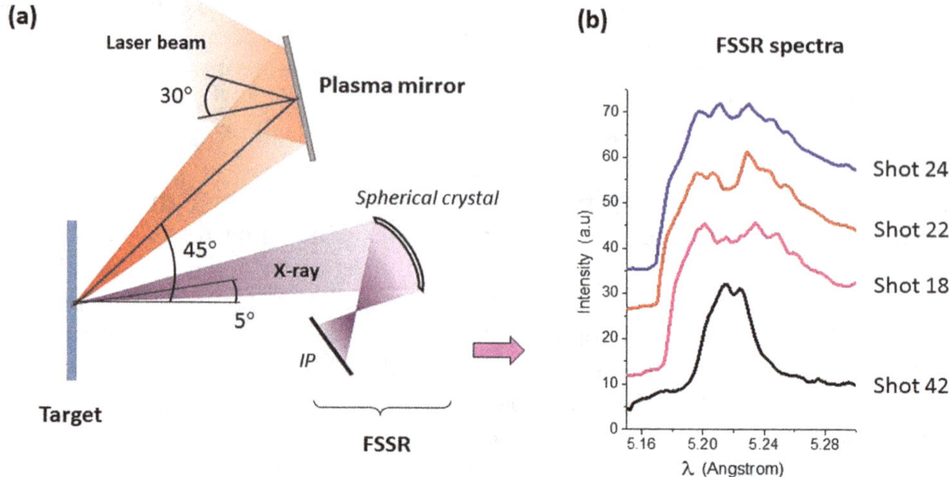

Figure 4.6. (a) Schematic of the experimental setup. (b) High resolution spectral measurements of the Ly$_\beta$ line of Si XIV obtained in different laser shots (see table 4.1 for details of the laser shot parameters) [53].

1054 nm and a pulse of the duration at the full-width-half-maximum (FWHM) of ~1.0 ps. In the previous experiments [1], the OPCPA technology was used, enabling an amplified spontaneous emission (ASE) to the peak-intensity contrast ratio exceeding 10^{-9}. In the experiments reported in paper [53] a plasma mirror was employed to increase the contrast ratio to the 10^{-11} range, ensuring that the main pulse interacted with an unperturbed, cold target. The highest laser pulse energy measured before the compressor was ~ 620 J, while the efficiency of the beam delivery line including the laser compressor, f/3 off-axis parabolic mirror and plasma mirror was measured as 50%, which corresponds to ~ 300 J laser energy on target on average. The p-polarized, focusing beam was reflected by the plasma mirror, to a focal spot of 7 μm (FWHM), at an incident angle of 45° to the normal to the target surface, as shown in figure 4.6. The central focal spot contained 30% of the laser energy, resulting in a maximum intensity of ~(2–3) × 10^{20} W cm^{-2}.

The x-ray emission of the plasma was registered by means of a focusing spectrometer, with a spatial resolution (FSSR), at the directions close to the normal to the target surface, from the rear side of the target. To obtain the spectra with a high spectral resolution ($\lambda/\delta\lambda \sim 3000$) in a rather broad range from 5.15 to 5.85 Å, the FSSR was equipped with quartz spherically bent (radius of curvature $R = 150$ mm) crystal 10–11 ($2d \sim 6.6$ A) installed at a distance of 332 mm from the target.

The spectra were recorded by means of Fujifilm TR Image Plate (IP) detectors protected against the exposure to the visible light by two layers of 1 μm-thick polypropylene $(C_3H_6)_n$ with 0.2 μm Al coating. Additionally, to prevent saturation of the detectors, the Mylar $(C_{10}H_8O_4)$ filter of 5 μm thickness was placed at the magnet entrance. The background fogging and crystal fluorescence due to intense fast electrons was limited by using a pair of 0.5 T neodymium–iron–boron permanent magnets that formed a 10 mm wide slit in front of each crystal.

Typical K-shell x-ray spectra of Si measured from the rear side of different solid targets are presented in figure 4.6(b).

Four types of targets were used in the experiment [53]. The parameters of the targets and laser pulses are presented in table 4.1. Some of Si targets were coated with or buried into thin plastic (CH) in order to keep the plasma of the main layer at solid densities regardless of the impact of the laser prepulse. The addition of the plastic layers on the front or both sides of the target is indicated in the first column of table 4.1, where the laser irradiated layer is shown first.

In shots A, B, C, the experimental profiles, indicated by 'Exp' in figure 4.7, exhibit two very distinct bump–dip–bump structures. The dips are located practically symmetrically with respect to the unperturbed wavelength of the line: one dip—in the blue wing, the other—in the red wing. Below we present the evidence that these are Langmuir-wave-induced dips (L-dips). Moreover, each of them is an L-super-dip, i.e. the superposition of two L-dips at the same location.

The Ly-beta line has two Stark components in each wing, corresponding to $q = 1$ and $q = 2$. Here $q = n_1 - n_2$ is the electric quantum number expressed via the parabolic quantum numbers n_1 and n_2: $q = 0, \pm 1, \pm 2, \ldots, \pm(n-1)$, where n is the principal quantum number (the electric quantum number marks Stark components of Ly-lines). Therefore, the L-dip in the profile of the component of $q = 1$ due to the two-quantum resonance ($s = 2$) coincides by its location with the L-dip in the profile of the component of $q = 2$ due to the one-quantum resonance ($s = 1$). The superposition of two different L-dips at the same location results in the L-super-dip with significantly enhanced visibility.

This L-super-dip is observed twice in the experimental profiles of the Si XIV Ly-beta line in shots A, B, and C: one—in the blue part and the other in the red part. These L-super-dips are located practically symmetrically at the distance $\Delta\lambda_{\mathrm{dip}}(N_e) = 24$ mÅ from the unperturbed wavelength. For $|q|s = 2$, this translates into the electron density $N_e = 2.2 \times 10^{22}$ cm^{-3}. (We remind the reader that this passive spectroscopic method for measuring N_e is just as accurate as the active spectroscopic method using the Thompson scattering, as was shown in the benchmark experiment [7].) This electron density is about the same as deduced in paper [34] by analyzing total intensities (rather than shapes) of Si XIV and Si XIII spectral lines (using code ATOMIC) in the plasma region (called zone 1 in paper [34]), from which these lines were emitted, the experiment being performed at the same laser facility at about the same incident laser intensities.

As for the would-be L-dip corresponding to $|q|s = 1$, i.e. the L-dip in the profile of the Stark component of $q = 1$ due to the one-quantum resonance ($s = 1$), it is not visible in the experimental profiles for the following reason. The location of the predicted L-dip is too close to the central part of the profile. The central part of the profile corresponds to the relatively small values of the field F, which are not quasistatic—see, for example, review [46]. Therefore, the would-be L-dips, corresponding to $|q|s = 1$ would not be observed.

At the absence of the low-frequency electrostatic plasma turbulence (LET), the quasistatic field required for the formation of the L-dips would be represented by the ion microfield. In this case, for relatively high electron densities, such as $N_e > 10^{22}$ cm^{-3}, due

Table 4.1. Target and laser shot parameters used in the experiments [53].

Target/Shot	Laser energy, J	Laser energy on target, J	Laser pulse length, ps	Focal spot diameter, μm	Intensity on the target, 10^{20} W cm^{-2}
Si 2 μm/Shot A	620 (±10%)	310	1.2	7	2.0
CH 2 μm + Si 2 μm + CH 2 μm/Shot B	490 (±10%)	245	1	7	1.9
CH 2 μm + Si 2 μm/Shot C	520 (±10%)	260	1.1	7	1.9
Si_3N_4 0.5 μm/Shot D	225(±10%)	110	1	7	0.9

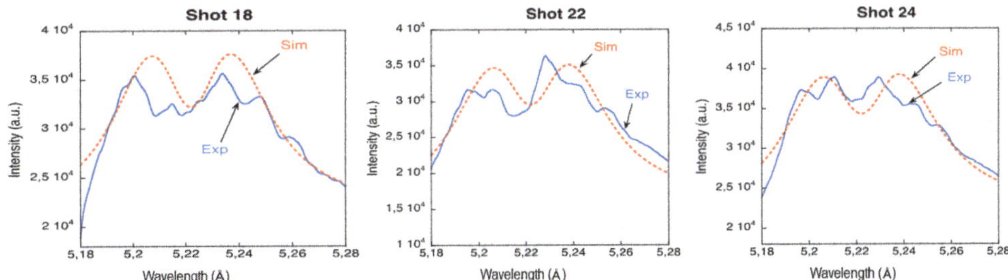

Figure 4.7. Experimental profiles of the Si XIV Ly-beta line in shots A, B, C (blue color in the online version) and their comparison with simulations using code FLYCHK (red color in the online version) at the electron density $N_e = 6 \times 10^{23}$ cm^{-3} and the temperature $T = 500$ eV [53].

to the spatial non-uniformity of the ion microfield the mid-point between the two L-dips in the pair would be significantly red-shifted [7, 8]. However, such a shift was not observed in the experiment presented in paper [53]. This was an indication that the quasistatic field required for the formation of the L-dips was represented primarily by the LET dominating the ion microfield.

Another indication of the presence of the LET comes from modeling the experimental profile using the code FLYCHK. This is an advanced code, but it does not take into account of the LET and the L-dips (the modelled profiles are shown in figure 4.7). It yielded $T = 500$ eV and $N_e = 6 \times 10^{23}$ cm^{-3}. This value of N_e is one and a half orders of magnitude higher than the electron density $N_e = 2.2 \times 10^{22}$ cm^{-3} deduced from the experimental L-dips.

The physical mechanism producing simultaneously the Langmuir waves and the LET (specifically, the ion acoustic turbulence) out of the laser field is the parametric decay instability (PDI). PDI is a nonlinear process, in which an electromagnetic wave decays into a Langmuir wave and an ion acoustic wave at the surface of the critical density N_c determined from the equation

$$\omega = \omega_{pe}(N_c), \tag{4.4}$$

where ω is the laser frequency. For the laser frequency ω corresponding to the wavelength λ of approximately 1054 nm in the experiment [53], equation (4.4) yields $N_c = 1.0 \times 10^{21}$ cm^{-3}. However, for relativistic laser intensities $I > 10^{18}$ W cm^{-2}, i.e. those corresponding to the present experiment, the 'relativistic' critical electron density N_{cr} becomes greater than N_c, as discussed in section 4.1. For the linearly-polarized laser radiation, it becomes N_c given by equation (4.3). For the laser intensities of the incident laser wave at the surface of the target $I_{surf} = 2 \times 10^{20}$ W cm^{-2}, used in shots A, B and C, equation (4.3) yields $N_{cr} = 1 \times 10^{22}$ cm^{-3}, which is just a factor of two smaller than the electron density $N_e = 2.2 \times 10^{22}$ cm^{-3} deduced from the experimental L-dips.

However, the actual intensity of the transverse electromagnetic wave in the plasma can be significantly greater than the intensity of the incident laser radiation at the surface of the target because of a number of physical effects. One of them is the self-focusing of the laser beam in plasmas—see, e.g. papers [40, 42] and reviews

[22, 43] providing details of the process of the laser propagation in the plasma corona at the overcritical density. The other relevant effects enhancing the transverse electromagnetic wave in the plasma are Raman and Brillouin backscattering. There are experimental proofs of such enhancement—see, e.g. paper [44]. In the experiment presented in paper [53] in shots A, B and C, for the relativistic critical density to be approximately equal to the density $N_e = 2.2 \times 10^{22}$ cm^{-3} deduced by the spectroscopic analysis, would require only a factor of two enhancement in the laser amplitude.

Thus, it was the PDI at the surface of the relativistic critical density that produced simultaneously the Langmuir waves and the ion acoustic turbulence in shots A, B and C. The amplitude E_0 of the Langmuir wave can be determined immediately from the experimental profile by using the expression for the half-width $\delta\lambda_{1/2}$ of the L-dip, i.e. the separation between the dip and the nearest 'bump' given by equation (4.2). Substituting the experimental $\delta\lambda_{1/2}$ in equation (4.2), the authors of paper [54] obtained the following values of the amplitude of the Langmuir waves: $E_0 = 0.7$ GV cm^{-1}, 0.5 GV cm^{-1}, and 0.6 GV cm^{-1} for shots A, B and C, respectively.

The resonant value of the quasistatic field F_{res} responsible for the formation of the L-dips, can be determined from the resonance condition

$$\omega_F(F_{\text{res}}) = s\omega_{\text{pe}}(N_e). \tag{4.5}$$

where ω_F is the Stark splitting of hydrogenic energy levels, caused by a quasistatic field $\mathbf{F}$ in a plasma: $\omega_F = 3n\hbar F/(2Z_r m_e e)$. For the formation of the L-dips the value of F_{res} is required to be at least several times higher than E_0 (the L-dips cannot form if $E_0/F_{\text{res}} > 0.5$)—see [8]. For shots A, B and C, equation (4.5) yields: $F_{\text{res}} = 6.5$ GV cm^{-1} for $s = 2$ and $F_{\text{res}} = 3.25$ GV cm^{-1} for $s = 1$. These values of F_{res} are about 10 and 5 times higher than the Langmuir wave amplitude E_0, respectively. Thus, the condition necessary for the formation of the L-dips was fulfilled in shots A, B and C.

For a detailed quantitative analysis/modeling, the authors of paper [53] calculated the theoretical profiles, providing the best fit to the experimental profiles from shots A, B and C as follows. The total quasistatic field $\mathbf{F}$ is the vector sum of two contributions: $\mathbf{F} = \mathbf{F}_t + \mathbf{F}_i$. The first contribution $\mathbf{F}_t$ is the field of a LET, while the second contribution $\mathbf{F}_i$ is the quasistatic part of the ion microfield. The authors of paper [54] employed the results from paper [48] to calculate the distributions of the total quasistatic field $\mathbf{F} = \mathbf{F}_t + \mathbf{F}_i$. Specifically, they calculated the distribution of the total quasistatic field $\mathbf{F}$ in the form of the convolution of the APEX distribution of $\mathbf{F}_i$ [49] with the Rayleigh-type distribution of $\mathbf{F}_t$ [47]. (It should be noted that for the case where the characteristic value of the LET is much greater than the characteristic value of the ion microfield, the analytical results from paper [54] provide a robust way to calculate the distribution of the total quasistatic field without calculating the convolution, though the authors of paper [53] did not use these results.) The authors of paper [53] also took into account the broadening by the electron microfield, the Doppler, the instrumental broadenings, as well as the theoretically expected asymmetry of the profiles (for the theory of the asymmetry they referred to papers [55, 56] and references therein). For calculating the details of the spectral line shape

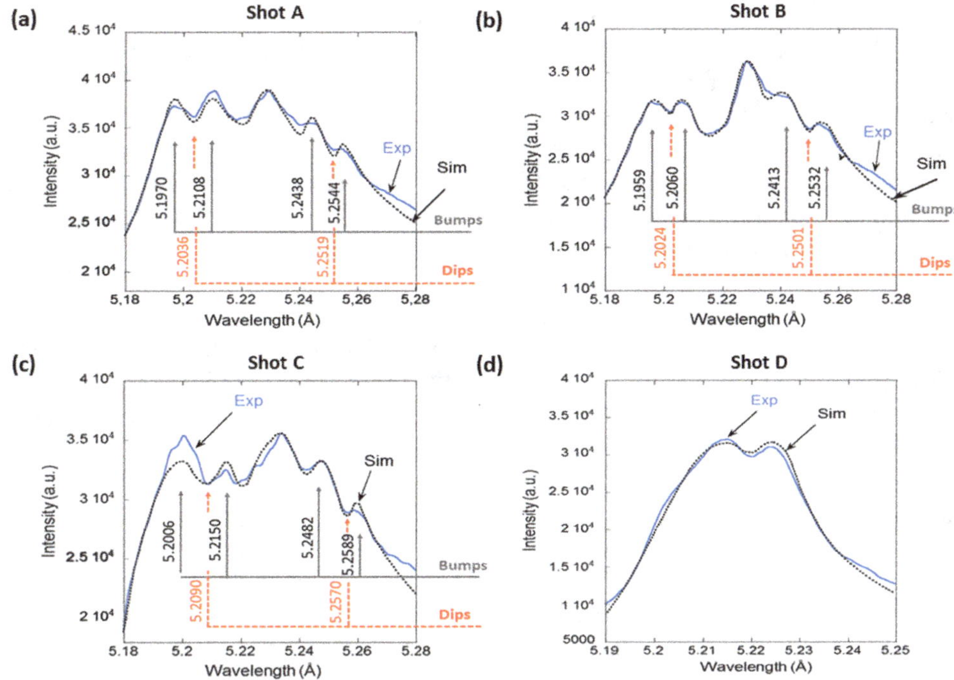

Figure 4.8. Comparison of the experimental profiles of the Si XIV Ly-beta line (solid line, blue in the online version, marked Exp) with the theoretical profiles (dotted line, black, marked Sim) allowing for the effects of the Langmuir waves, the LET, and all other broadening mechanisms (see the text) [53]. In the profiles from shots A, B and C, there are clearly seen 'bump–dip–bump' structures (both in the red and blue parts of the profiles) typical for the L-dips phenomenon. The following parameters provided the best fit: (a) $N_e = 2.2 \times 10^{22}$ cm^{-3}, $T = 600$ eV, $F_{t,\mathrm{rms}} = 4.8$ GV cm^{-1}, $E_0 = 0.7$ GV cm^{-1}; (b) $N_e = 2.2 \times 10^{22}$ cm^{-3}, $T = 550$ eV, $F_{t,\mathrm{rms}} = 4.4$ GV cm^{-1}, $E_0 = 0.5$ GV cm^{-1}; (c) $N_e = 2.2 \times 10^{22}$ cm^{-3}, $T = 600$ eV, $F_{t,\mathrm{rms}} = 4.9$ GV cm^{-1}, $E_0 = 0.6$ GV cm^{-1}; (d) $N_e = 6.6 \times 10^{21}$ cm^{-3}, $T = 550$ eV, $F_{t,\mathrm{rms}} = 2.0$ GV cm^{-1}, $E_0 = 2.0$ GV cm^{-1}.

in the regions of L-dips they employed the analytical solution from references [8, 50] for the wave functions of the quasienergy states, the latter being caused simultaneously by all harmonics of the total electric field $\mathbf{E}(t) = \mathbf{F} + \mathbf{E}_0 \cos(\omega_{\mathrm{pe}}t)$ (vectors $\mathbf{F}$ and $\mathbf{E}_0$ are not collinear).

Figure 4.8 shows the comparison of the theoretical profiles, allowing, in particular, for the LET and L-dips, with the corresponding experimental profiles from shots A, B and C. The theoretical profiles were calculated at $N_e = 2.2 \times 10^{22}$ cm^{-3} and the temperature $T = 600$ eV, 550 eV, and 600 eV for shots 18, 22, and 24, respectively. The comparison demonstrates good agreement between the theoretical and experimental profiles, and thus reinforces the interpretation of the experimental profiles.

This modeling yielded the following values of the root-mean-square field of the LET: $F_{t,\mathrm{rms}} = 4.8$ GV cm^{-1}, 4.4 GV cm^{-1}, and 4.9 GV cm^{-1} for shots A, B and C respectively. For comparison, the characteristic ion microfield $F_{i,\mathrm{typ}} = 2.603\ eZ^{1/3}N_e^{2/3}$ was 1.5 GV cm^{-1}.

Then the authors of paper [24] proceeded to analyzing the experimental profile of the Si XIV Ly-beta line in shot D—see figure 4.8(d). The experimental profile does not show bump–dip–bump structures—in distinction to shots A, B and C. In shot D the incident laser intensity was $I = 8.8 \times 10^{19}$ W cm^{-2}, i.e. it was significantly lower than in shots A, B and C. The corresponding relativistic critical density is $N_e = 6.6 \times 10^{21}$ cm^{-3}. The modeling using code FLYCHK (that does not take into account the LET and the Langmuir waves) yielded $N_e = 1.7 \times 10^{23}$ cm^{-3}, which is one and a half orders of magnitude higher than the relativistic critical density.

It is unlikely that the experimental profile in shot D, produced at the significantly smaller laser intensity than in shots A, B and C, would be emitted from the region of the electron density $N_e = 1.7 \times 10^{23}$ cm^{-3} by an order of magnitude higher than the region of the electron density $N_e = 2.2 \times 10^{22}$ cm^{-3}, from which the experimental profiles were emitted in shots A, B and C. The most probable interpretation of the experimental profile in shot D is the following.

In shot D the electron density was significantly lower than in shots A, B and C. Therefore, the damping of the Langmuir waves was significantly lower, which could allow the Langmuir waves to reach a significantly higher amplitude. Figure 4.8(d) shows the comparison of the experimental profile from shot D with the modeling based on the code allowing for the LET and the Langmuir waves at $N_e = 6.6 \times 10^{21}$ cm^{-3}, $T = 550$ eV, $F_{t,\mathrm{rms}} = 2.0\,\mathrm{GV\,cm}^{-1}$, $E_0 = 2.0\,\mathrm{GV\,cm}^{-1}$. It is seen that this theoretical profile is in good agreement with the experimental profile and it does not exhibit bump–dip–bump structures.

The authors of paper [53] provided further details on the reason for this good agreement. When the ratio $E_0/F_{\mathrm{res}} > 0.5$, the L-dips cannot form—as mentioned above with reference to book [8]. The determination of the resonant value F_{res} of the quasistatic field from equation (4.5) is valid only for $E_0 \ll F_{\mathrm{res}}$. For being valid for arbitrary ratios of E_0 to F_{res}, equation (4.5) should be modified as follows (according to book [8]):

$$\Omega_F(F_{\mathrm{res}})g(\varepsilon) = s\omega_{\mathrm{pe}}(N_e), \quad g(\varepsilon) = (1 + \varepsilon^2)^{1/2}\mathrm{EllipticE}[\varepsilon/(1 + \varepsilon^2)^{1/2}], \quad \varepsilon = E_0/F_{\mathrm{res}}, \quad (4.6)$$

where $\omega_F = 3n\hbar F/(2Z_r m_e e)$ and EllipticE[...] is the complete elliptic integral of the second kind. For $N_e = 6.6 \times 10^{21}$ cm^{-3} and $E_0 = 2.0\,\mathrm{GV\,cm}^{-1}$, equation (4.6) yields $F_{\mathrm{res}} = 1.7\,\mathrm{GV\,cm}^{-1}$ for $s = 1$ (so that $E_0/F_{\mathrm{res}} = 1.2$) and $F_{\mathrm{res}} = 3.5\,\mathrm{GV\,cm}^{-1}$ for $s = 2$ (so that $E_0/F_{\mathrm{res}} = 0.6$). Thus, both for the one-quantum resonance ($s = 1$) and for the two-quantum resonance ($s = 2$), one gets $E_0/F_{\mathrm{res}} > 0.5$, so that the L-dips were not able to form. So, most probably in shot D the PDI did occur at the surface of the relativistic critical density, resulting in the development of both the Langmuir waves and the ion acoustic turbulence, but without producing the L-dips in the experimental line profile. It should be noted that, while for shots A, B and C the electron density was unequivocally deduced simply from the locations of the L-dips, in shot D it turned out possible to deduce the electron density by modeling the entire experimental profile using the code that allows for the interplay of the LET and the Langmuir waves.

In summary, the authors of paper [53] performed an in-depth spectroscopic study of the simultaneous production of Langmuir waves and of the ion acoustic turbulence at the surface of the relativistic critical density under the relativistic laser-plasma interaction. They demonstrated a reliable reproducibility of the L-dips at the same locations in the experimental profiles, as well as of the deduced parameters (fields) of the Langmuir waves and ion acoustic turbulence. By doing so, they expanded applications of the intra-Stark spectroscopy and reinforced the validity of the physics behind these different waves as being caused by the parametric decay instability at the surface of the relativistic critical density.

The authors of paper [53] also employed, for the first time, the most rigorous condition of the dynamic resonance, on which the intra-Stark spectroscopy is based, compared to all previous studies in all kinds of plasmas in a wide range of electron densities. As a result, they were able to show how different interplay between the Langmuir wave field with the field of the ion acoustic turbulence led to distinct manifestations in the spectral line profiles, including the disappearance of the L-dips.

The results presented in paper [53] should motivate further applications of the intra-Stark spectroscopy to studies of laser-produced plasmas in general, and to studies of relativistic laser-plasma interactions in particular.

References

[1] Oks E *et al* 2017 *Opt. Express* **25** 1958–72

[2] Tsytovich V N 1977 *Theory of Turbulent Plasmas* (Berlin: Springer)

[3] Kadomtsev B B 1965 *Plasma Turbulence* (Cambridge, MA: Academic)

[4] Zhuzhunashvili A I and Oks E 1977 *Sov. Phys. JETP* **46** 1122

[5] Oks E and Rantsev-Kartinov V A 1980 *Sov. Phys. JETP* **52** 50

[6] Jian L, Shali X, Qingguo Y, Lifeng L and Yufen W 2013 *J. Quant. Spectrosc. Radiat. Transf.* **116** 41

[7] Oks E, Böddeker S and Kunze H J 1991 *Phys. Rev.* A **44** 8338

[8] Oks E 1995 *Plasma Spectroscopy: The Influence of Microwave and Laser Fields Springer Series on Atoms and Plasmas* vol 9 (Berlin: Springer)

[9] Ebeling W 1984 *Transport Properties of Dense Plasmas* (Basel: Birkhäuser)

[10] Manfredi G and Dendy R O 1997 *Phys. Plasmas* **4** 628

[11] Antonov A S, Zinov'ev O A, Rusanov V D and Titov A V 1970 *Sov. Phys. JETP* **31** 838

[12] Zagorodnikov S P, Smolkin G E, Striganova E A and Sholin G V 1971 *Sov. Phys. Doklady* **15** 1122

[13] Zavojskij E K, Kalinin J G, Skorjupin V A, Shapkin V V and Sholin G V 1971 *Sov. Phys. Doklady* **15** 823

[14] Levine M A and Gallagher C C 1970 Stark broadening for turbulence studies in a confined plasma *Phys. Lett.* A **32** 14

[15] Ben-Yosef N and Rubin A G 1970 *Phys. Lett.* A **33** 222

[16] Berezin A B, Dubovoj A V and Ljublin B V 1972 *Sov. Phys. Tech. Phys.* **16** 1844

[17] Babykin M V, Zhuzhunashvili A I, Oks E, Shapkin V V and Sholin G V 1974 *Sov. Phys. JETP* **38** 86

[18] Volkov J F, Djatlov V G and Mitina A I 1975 *Sov. Phys. Tech. Phys.* **19** 905

[19] Berezin A B, Ljublin B V and Jakovlev D G 1983 *Sov. Phys. Tech. Phys.* **28** 407

[20] Koval A N and Oks E 1983 *Bull. Crimean Astrophys. Observ.* **67** 78

[21] Frank A G, Gavrilenko V P, Kyrie N P and Oks E 2006 *J. Phys. B: At. Mol. Opt. Phys.* **39** 5119

[22] Pukhov A 2003 *Rep. Prog. Phys.* **66** 47

[23] McKenna P, Neely D, Bingham R and Jaroszynski D (ed) 2013 *Laser-Plasma Interactions and Applications* (Berlin: Springer)

[24] Renner O, Dalimier E, Oks E, Krasniqi F, Dufour E, Schott R and Foerster E 2006 *J. Quant. Spectrosc. Radiat. Transf.* **99** 439

[25] Gangadhara R T and Krishan V 1990 Absorption of electromagnetic waves in astrophysical plasmas *Basic Plasma Processes on the Sun* ed E R Priest and V Krishan (Berlin: Springer)

[26] Frank A G 2010 *Phys. – Uspekhi* **53** 941

[27] Bulanov S V, Dogie V A and Frank A G 1984 *Phys. Scr.* **29** 66

[28] Dromey B *et al* 2007 *Phys. Rev. Lett.* **99** 085001

[29] Danson C N *et al* 2004 *Nucl. Fusion* **44** 5239

[30] Musgrave I, Shaikh W, Galimberti M, Boyle A, Hernandez-Gomez C, Lancaster K and Heathcote R 2010 *Appl. Opt.* **49** 6558

[31] Faenov A Y, Pikuz S A, Erko A I, Bryunetkin B A, Dyakin V M, Ivanenkov G V, Mingaleev A R, Pikuz T A, Romanova V M and Shelkovenko T A 1994 *Phys. Scr.* **50** 333

[32] Colgan J *et al* 2013 *Phys. Rev. Lett.* **110** 125001

[33] Pikuz S A *et al* 2013 *High Energy Density Phys.* **9** 560

[34] Colgan J *et al* 2016 *Europhys. Lett.* **114** 35001

[35] Maine P, Strickland D, Bado P, Pessot M and Mourou G 1988 *IEEE J. Quantum Electron.* **24** 398

[36] Akhiezer A I and Polovin R V 1956 *Sov. Phys. JETP* **3** 696

[37] Lünow W 1968 *Plasma Phys.* **10** 879

[38] Guerin S, Mora P, Adam J C, Heron A and Laval G 1996 *Phys. Plasmas* **3** 2693

[39] Gray R J *et al* 2014 *New J. Phys.* **16** 113075

[40] Fuchs J *et al* 1998 *Phys. Rev. Lett.* **80** 2326

[41] Chung H K, Chen M H, Morgan W L, Ralchenko Y and Lee R W 2005 *High Energy Density Phys.* **1** 3

[42] Pukhov A and Meyer-ter-Vehn J 1997 *Phys. Rev. Lett.* **79** 2686

[43] Mourou G A, Tajima T and Bulanov S V 2006 *Rev. Mod. Phys.* **78** 309

[44] Sauvan P, Dalimier E, Oks E, Renner O, Weber S and Riconda C 2009 *J. Phys. B: At. Mol. Opt. Phys.* **42** 195501

[45] Gibbon P 2003 *Short Pulse Laser Interaction with Matter* (London: Imperial College Press)

[46] Lisitsa V S 1977 *Sov. Phys. Uspekhi* **122** 449

[47] Oks E and Sholin G V 1976 *Sov. Phys. Tech. Phys.* **21** 144

[48] Oks E and Dalimier E 2011 *Int. Rev. At. Mol. Phys.* **2** 43

[49] Iglesias C A, Dewitt H E, Lebowitz J L, MacGowan D and Hubbard W B 1985 *Phys. Rev. A* **31** 1698

[50] Gavrilenko V P and Oks E A 1981 *Sov. Phys. JETP* **53** 1122

[51] Lichters R, Meyer-ter-Vehn J and Pukhov A 1996 *Phys. Plasmas* **3** 3425

[52] Kruer W L 2003 *The Physics of Laser Plasma Interactions* (Boulder, CO: Westview)

[53] Oks E *et al* 2017 *J. Phys. B: At. Mol. Opt. Phys.* **50** 245006

[54] Dalimier E and Oks E 2017 *J. Phys. B: At. Mol. Opt. Phys.* **50** 025701

[55] Djurovic S, Ćirišan M, Demura A V, Demchenko G V, Nikolić D, Gigosos M A and González M A 2009 *Phys. Rev. E* **79** 046402

[56] Demura A V, Demchenko G V and Nikolic D 2008 *Eur. Phys. J. D* **46** 111

IOP Publishing

Advances in X-Ray Spectroscopy of Laser Plasmas

Eugene Oks

Chapter 5

Role of ultra-intense magnetic fields in laser plasma spectroscopy

GigaGauss (GG) and even multi-GG magnetic fields are expected to be developed during relativistic laser–plasma interactions: these fields should be localized at the surface of the relativistic critical density—see, e.g. review [1] and references therein.

(We remind readers that 'relativistic laser–plasma interactions' relate to phenomena arising when plasma electrons reach relativistic velocities under a highly-intense laser field, such as, for example the increase of the electron mass resulting in the decrease of the plasma electron frequency.) In particular, according to equation (11) from paper [2], the maximum magnetic field B_{max} is related to the laser intensity I as follows:

$$B_{max}(G) = 10^{-1}[I \ (Wcm^{-2})]^{1/2}. \tag{5.1}$$

In recent experiments (see, e.g. paper [3, 4]), laser intensities $I \sim 10^{21}$ W cm^{-2} have been achieved. Therefore, according to equation (5.1), the magnetic fields can be as high as $B_{max} \sim 3$ GG.

On the experimental side, in paper [5] magnetic fields $B \sim 0.7$ GG were measured by using the polarization measurements (the Cotton–Mouton effect of an induced ellipticity) of high-order VUV laser harmonics generated at the incident irradiation intensity $I = 9 \times 10^{19}$ W cm^{-2}. In an earlier experiment [6, 7], magnetic fields up to $B \sim 0.4$ GG were measured at the incident irradiation intensity up to $I = 9 \times 10^{19}$ W cm^{-2}, by a method also using the self-generated harmonics of the laser frequency and the fact that the magnetized plasma is birefringent (the Cotton–Mouton effect) and/or optically active (the Faraday effect of the rotation of the polarization vector) depending on the propagation direction of the electromagnetic wave.

In paper [8], which we follow below, the authors proposed a method for measuring GG magnetic fields based on the phenomenon of Langmuir-wave-cased dips (L-dips) in x-ray line profiles. The L-dips were observed in several experimental

doi:10.1088/978-0-7503-3375-7ch5

© IOP Publishing Ltd 2020

spectroscopic studies of relativistic laser–plasma interactions, as presented in chapter 4 of this book. The corresponding theory is briefly presented in chapter 3 of the current book.

Generally, there could be two sets of L-dips in the spectral line profile at distances $\Delta\omega_{\text{dip}}$ from the unperturbed frequency ω_0 of the spectral line. One set, located at

$$\Delta\omega_{\text{dip}}{}^{(\alpha)} = (q_\alpha - q_\beta n_\beta/n_\alpha)s\omega_{\text{pe}} \qquad (5.2)$$

results from the resonance with the splitting of the upper sublevel α (of the principal quantum number n_α) involved in the radiative transition. Another set located at

$$\Delta\omega_{\text{dip}}{}^{(\beta)} = (q_\alpha n_\alpha/n_\beta - q_\beta)s\omega_{\text{pe}} \qquad (5.3)$$

results from the resonance with the splitting of the lower sublevel β (of the principal quantum number n_β) involved in the radiative transition. Here $q = n_1 - n_2$ is the electric quantum number expressed via the parabolic quantum numbers n_1 and n_2: $q = 0, \pm1, \pm2, \ldots, \pm(n - 1)$. The electric quantum number marks Stark components of hydrogenic spectral lines. It should be emphasized that for the Ly-lines, there is no second set of the L-dips at $\Delta\omega_{\text{dip}}{}^{(\beta)}$ because there is no linear Stark splitting of the state of $n = 1$. Below for brevity the subscript 'pe' has been omitted and there is used ω instead of ω_{pe}.

In paper [9], for the specific case of the one-quantum resonance ($s = 1$) in hydrogen atoms ($Z_r = 1$), Gavrilenko generalized equations (5.2) and (5.3) for the situation where there is also a magnetic field $\mathbf{B}$ in plasmas. His corresponding formulas are as follows:

$$\Delta\omega_{\text{dip}}{}^{(\alpha)} = \omega\{(n' + n'')_\alpha - [(n' + n'')_\beta/n_\alpha][(n_\alpha{}^2 - n_\beta{}^2)b_0{}^2 + n_\beta{}^2]^{1/2}\}, \qquad (5.4)$$

$$\Delta\omega_{\text{dip}}{}^{(\beta)} = \omega\left\{[(n' + n'')_\alpha/n_\beta]\left[n_\alpha{}^2 - \left(n_\alpha{}^2 - n_\beta{}^2\right)b_0{}^2\right]^{1/2} - (n' + n'')_\beta\right\}. \qquad (5.5)$$

Here the quantum numbers n' and n'' correspond to the basis of the wave functions diagonalizing the Hamiltonian of a hydrogen atom in a non-collinear static electric ($\mathbf{F}$) and magnetic ($\mathbf{B}$) fields (see, e.g. paper [10]):

$$n', n'' = -j, -j + 1, \ldots, j; \qquad j = (n - 1)/2. \qquad (5.6)$$

The quantity b_0 in equations (5.4) and (5.5) is the scaled, dimensionless magnetic field

$$b_0 = \mu_0 B/(\hbar\omega), \qquad (5.7)$$

where μ_0 is the Bohr magneton.

The authors of paper [8] further slightly generalized Gavrilenko's formulas by allowing for any number of quanta s involved in the resonance and for any nuclear charge Z_r of hydrogenic atoms/ions:

$$\Delta\omega_{\text{dip}}{}^{(\alpha)} = s\omega\left\{(n' + n'')_\alpha - \left[(n' + n'')_\beta/n_\alpha\right]\left[(n_\alpha{}^2 - n_\beta{}^2)b^2 + n_\beta{}^2\right]^{1/2}\right\}, \qquad (5.8)$$

$$\Delta\omega_{\mathrm{dip}}{}^{(\beta)} = s\omega\left\{\left[(n' + n'')_\alpha/n_\beta\right]\left[n_\alpha{}^2 - \left(n_\alpha{}^2 - n_\beta{}^2\right)b^2\right]^{1/2} - (n' + n'')_\beta\right\}, \qquad (5.9)$$

where the scaled dimensionless magnetic field b now reads:

$$b = \mu_0 B/(s\hbar\omega) = (1/s)[B\,(\mathrm{GG})/0.201][\omega(s^{-1})/(1.77 \times 10^{15})]^{-1}. \qquad (5.10)$$

For example, for the one-quantum resonance ($s = 1$), for the frequency $\omega = 1.77 \times 10^{15}$ s^{-1}, which is the frequency of the laser used, e.g. in experiments [3, 4], the quantity b reaches unity at $B = 0.201$ GG. It should be noted that the nuclear charge Z_r does not enter equations (5.8), (5.9), but obviously does affect the unperturbed frequency of the spectral line.

Here is the idea of a new method for measuring the magnetic fields. It is possible to select such a pair of the L-dip at $\Delta\omega_{\mathrm{dip}}{}^{(\alpha)}$ and the L-dip at $\Delta\omega_{\mathrm{dip}}{}^{(\beta)}$, both corresponding to the same combination of the sums $(n' + n'')_\alpha$ and $(n' + n'')_\beta$, such that the location of one of the two L-dips is unaffected by the magnetic field while the location of the other of the two L-dips is shifted by the magnetic field. Then from the relative separation of the two L-dips it is possible to determine the magnetic field.

Namely, this is about the following pairs of the L-dips. One pair corresponds to

$$(n' + n'')_\alpha = 0, \ (n' + n'')_\beta = -1, \qquad (5.11)$$

while another pair corresponds to

$$(n' + n'')_\alpha = 1, \ (n' + n'')_\beta = 0. \qquad (5.12)$$

The ratio

$$\Delta\omega_{\mathrm{dip}}{}^{(\alpha)}/\Delta\omega_{\mathrm{dip}}{}^{(\beta)} = (1/n_\alpha)[(n_\alpha{}^2 - n_\beta{}^2)b^2 + n_\beta{}^2]^{1/2} \qquad (5.13)$$

in the first case and the ratio

$$\Delta\omega_{\mathrm{dip}}{}^{(\beta)}/\Delta\omega_{\mathrm{dip}}{}^{(\alpha)} = \left(1/n_\beta\right)\left[n_\alpha{}^2 - \left(n_\alpha{}^2 - n_\beta{}^2\right)b^2\right]^{1/2} \qquad (5.14)$$

in the second case are simple functions of the magnetic field, as is seen from the above formulas.

Figure 5.1 shows the ratio $\Delta\omega_{\mathrm{dip}}{}^{(\alpha)}/\Delta\omega_{\mathrm{dip}}{}^{(\beta)}$ in the pair of the L-dips corresponding to $(n' + n'')_\alpha = 0$, $(n' + n'')_\beta = -1$, versus the scaled dimensionless magnetic field b for the Balmer-alpha line (solid curve) and for the Balmer-beta line (dashed curve).

It is seen that in the range of b presented in figure 5.1, the magnetic field significantly affects the relative positions of the L-dips, so that by measuring the latter it is possible to determine the magnetic field. For the laser frequency $\omega = 1.77 \times 10^{15}$ s^{-1} used, e.g. in experiments [3, 4], the range of $b \sim (1\text{--}10)$ corresponds to the range of the magnetic field $B \sim (0.2\text{--}2)$ GG for the one-quantum resonance and to $B \sim (0.4\text{--}4)$ GG for the two-quantum resonance. For $b \gg 10$, the possible L-dips at $\Delta\omega_{\mathrm{dip}}{}^{(\alpha)}$ would be shifted too far into the wings of the spectral lines, so that most probably they could not be observed.

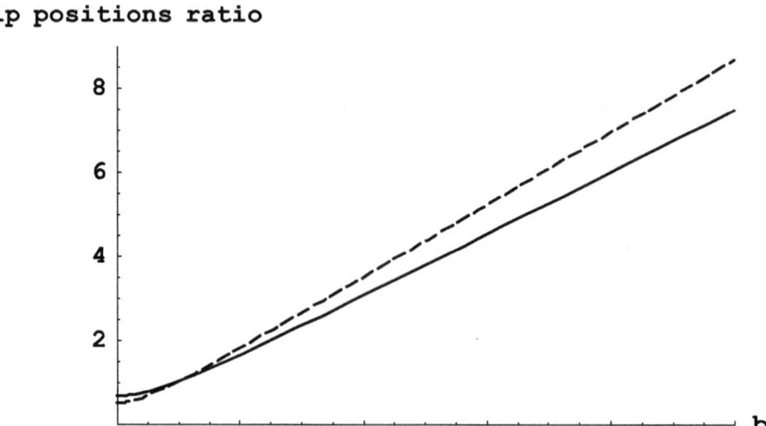

Figure 5.1. The ratio of positions $\Delta\omega_{\mathrm{dip}}^{(\alpha)}/\Delta\omega_{\mathrm{dip}}^{(\beta)}$ in the pair of the L-dips corresponding to $(n' + n'')_\alpha = 0$, $(n' + n'')_\beta = -1$, versus the scaled (dimensionless) magnetic field b (defined by equation (5.12)) for the Balmer-alpha line (solid curve) and for the Balmer-beta line (dashed curve) [8].

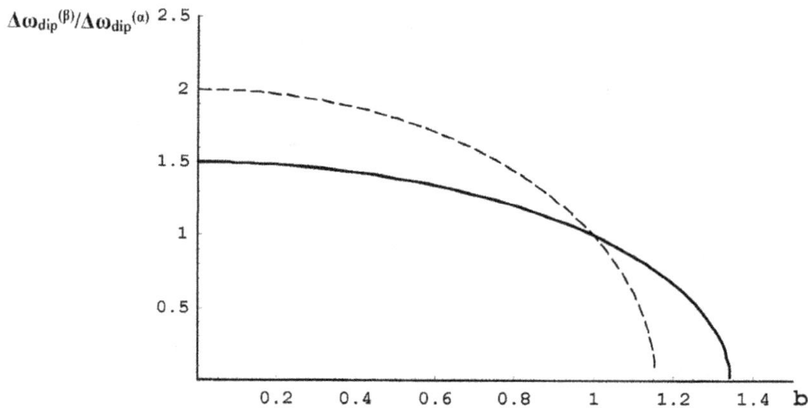

Figure 5.2. The ratio of positions $\Delta\omega_{\mathrm{dip}}^{(\beta)}/\Delta\omega_{\mathrm{dip}}^{(\alpha)}$ in the pair of the L-dips corresponding to $(n' + n'')_\alpha = 1$, $(n' + n'')_\beta = 0$, versus the scaled (dimensionless) magnetic field b (defined by equation (5.9)) for the Balmer-alpha line (solid curve) and for the Balmer-beta line (dashed curve). © 1998 IEEE. Reprinted, with permission, from [12].

Figure 5.2 presents the ratio $\Delta\omega_{\mathrm{dip}}^{(\beta)}/\Delta\omega_{\mathrm{dip}}^{(\alpha)}$ in the pair of the L-dips corresponding to $(n' + n'')_\alpha = 1$, $(n' + n'')_\beta = 0$, versus the scaled dimensionless magnetic field b for the Balmer-alpha line (solid curve) and for the Balmer-beta line (dashed curve).

For the laser frequency $\omega = 1.77 \times 10^{15}$ s^{-1} used, e.g. in experiments [3, 4], the range of b up to 1.4 corresponds to the magnetic field up to 0.3 GG for the one-quantum resonance and to the magnetic field up to 0.6 GG for the two-quantum resonance. It should be noted that if one would use the pair of the L-dips in the profiles of Stark components characterized by the quantum numbers from equation

(5.12), then according to equation (5.14) the range of b would be limited to $b_{\max} = n_\alpha^2/(n_\alpha^2 - n_\beta^2)$. This is because at $b_{\max} = n_\alpha^2/(n_\alpha^2 - n_\beta^2)$, the possible L-dips at $\Delta\omega_{\mathrm{dip}}^{(\beta)}$ would disappear.

Here is the following general formula for the ratio $\Delta\omega_{\mathrm{dip}}^{(\alpha)}/\Delta\omega_{\mathrm{dip}}^{(\beta)}$:

$$\Delta\omega_{\mathrm{dip}}^{(\alpha)}/\Delta\omega_{\mathrm{dip}}^{(\beta)} = \left\{ (n' + n'')_\alpha - \left[(n' + n'')_\beta/n_\alpha \right]\left[(n_\alpha^2 - n_\beta^2)b^2 + n_\beta^2 \right] \right\} \Big/$$
$$\left\{ \left[(n' + n'')_\alpha/n_\beta \right]\left[n_\alpha^2 - (n_\alpha^2 - n_\beta^2)b^2 \right]^{1/2} - (n' + n'')_\beta \right\} \tag{5.15}$$

Obviously the dependence on the scaled magnetic field b in the right side of equation (5.15) is more complicated than in the right side of equations (5.13) and (5.14). Nevertheless it seems to be still feasible to determine experimentally ultra-intense magnetic fields using any pair of the L-dips—as long as one dip in the pair is due to the resonance with the splitting of the upper level and the other dip in the pair is due to the resonance with the splitting of the lower level.

Here is some practical example based on measuring the relative shift of the L-dips in the profiles of the Balmer lines of Cu XXIX. (It is worth mentioning that it is technologically simple to make and use thin Cu foils to irradiate them by a powerful laser.) The wavelengths of the Balmer-alpha and Balmer-beta lines of Cu XXIX are 0.77 nm and 0.57 nm, respectively. This is practically the same range of the wavelength as was employed, e.g. in experiments [3, 4] while studying the L-dips in the profiles of the Ly-beta line of Si XIV and Al XIII. Therefore, the same kind of spectrometers can be used without any major additional tuning for experimental studies of possible L-dips in the profile of the Balmer-alpha and Balmer-beta lines of Cu XXIX, and thus for the experimental determination of GG (or sub-GG) magnetic fields.

The above diagnostic method can be implemented by using any hydrogenic spectral lines except the Lyman lines. The exception is due to the fact that for the Lyman lines, the lower level ($n = 1$) does not split. Consequently, there are no L-dips at $\Delta\omega_{\mathrm{dip}}^{(\beta)}$.

In paper [11] its authors suggested an alternative new method. This method can utilize any hydrogenic spectral lines, the Lyman lines being included. This new method is based on the effect of ultra-strong magnetic fields on the halfwidth of the L-dips (the halfwidth being best measurable as the separation between the primary minimum of the bump–dip–bump structure and the nearest bump). Below are some details.

The authors of paper [11] generalized the results from the appendix of Gavrilenko's paper [9] where he calculated the spectrum of the Ly-alpha line of hydrogen under non-orthogonal fields **B** and **F**. For presenting the results in the most universal form, they introduced the following three dimensionless quantities (in addition to the scaled dimensionless magnetic field b define by equation (5.10)):

$$f = \hbar F/(Z_r m_e e\omega), \qquad \gamma = \hbar E_0/(Z_r m_e e\omega), \qquad b_0 = \mu_0 B/(\hbar\omega). \tag{5.16}$$

In equation (5.16), f is the scaled dimensionless quasistatic electric field, γ is the scaled dimensionless amplitude of the Langmuir wave, and b_0 is the scaled magnetic field (scaled in a slightly different way compared to b from equation (5.10)).

Figure 5.3 shows the dependence of the scaled, dimensionless halfwidth $w_1 = \delta\omega_{\mathrm{dip}}^{(1)}/\omega$ of the L-dips in the Lyman-alpha line for the case of the one-quantum resonance versus the scaled magnetic field b_0 at $\gamma = 0.01$ for the following three values of the scaled quasistatic electric field: $f = 0.03$, $f = 0.1$, and $f = 0.3$. For the laser frequency $\omega = 1.77 \times 10^{15}\ \mathrm{s}^{-1}$ used, e.g. in experiments [3, 4], the range of b_0 up to 10 corresponds to the magnetic field B up to 2 GG.

From figure 5.3 one can see that at a fixed value of the quasistatic field F, the halfwidth of the L-dips decreases as the magnetic field increases. It is also seen that at a fixed value of the magnetic field B, the halfwidth of the L-dips increases as the quasistatic field F increases. It should be recalled that at $B = 0$ and $E_0 \ll F$, the halfwidth of the L-dips depended only on E_0, but did not depend on F. Thus, in strongly magnetized plasmas, the functional dependence of the halfwidth of the L-dips becomes more complicated: it depends not only on E_0, but also on the ratio B/F.

Figure 5.4 shows the same as figure 5.3, but for the case of the two-quantum resonance. One can see that in the case of the two-quantum resonance, the scaled halfwidth w_2 of the L-dips is a *non-monotonic function* of the scaled magnetic field b_0. This is a *counterintuitive result*. From the comparison of figures 5.3 and 5.4, it is also seen that under the two-quantum resonance, the halfwidth of the L-dips is significantly smaller than under the one-quantum resonance.

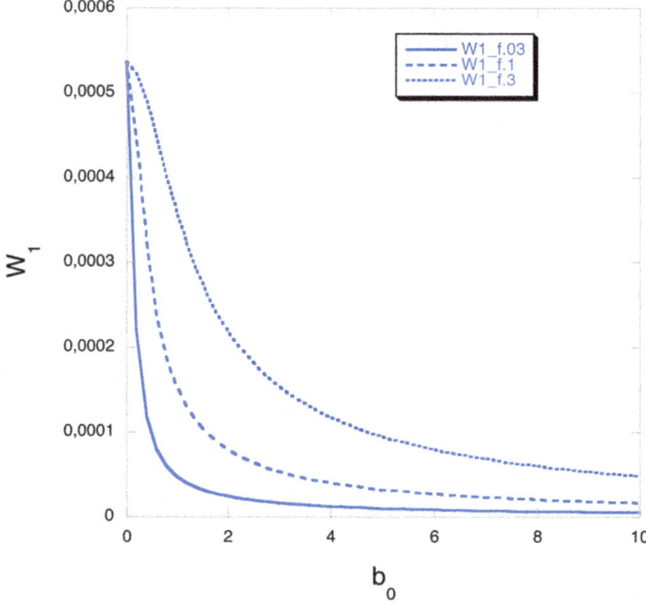

Figure 5.3. Dependence of the scaled halfwidth $w_1 = \delta\omega_{\mathrm{dip}}^{(1)}/\omega$ of the L-dips for the Lyman-alpha line in the case of the one-quantum resonance versus the scaled magnetic field b_0 (defined by equation (5.16)) at the scaled amplitude $\gamma = 0.01$ of the Langmuir wave (defined in equation (5.16)) for the following three values of the scaled quasistatic electric field (defined in equation (5.16)): $f = 0.03$ (solid line), $f = 0.1$ (dashed line), and $f = 0.3$ (dotted line). Reprinted from [11], copyright 2019, with permission from Elsevier.

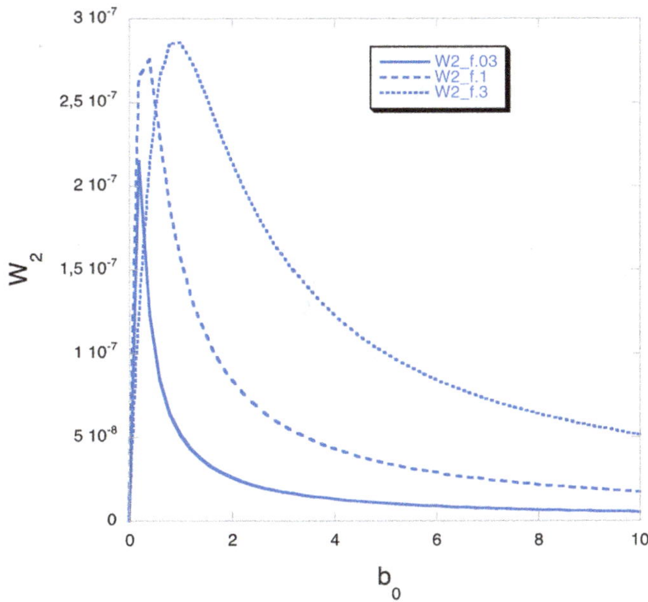

Figure 5.4. The same as in figure 5.3, but for the case of the two-quantum resonance. Reprinted from [11], copyright 2019, with permission from Elsevier.

It is important to emphasize that the utilization of the effect of the ultra-strong magnetic fields on the width of the L-dips allows measuring such fields by observing x-ray Lyman lines, which are routinely studied in laser–plasma interaction experiments. This is a great advantage compared to the new method from paper [8] (presented in the first part of this chapter) where the authors proposed employing the effect of such fields on the separation of the L-dips from one another. Namely, the second new method [11] is advantageous because it is not limited to using non-Lyman lines.

In summary, ultrastrong magnetic fields affect both the separation of the L-dips from one another and the halfwidth of the L-dips. Either one of these effects can be used to measure sub-GG and GG magnetic fields. Earlier there was proposed another diagnostic of magnetic fields in plasmas based on the polarization measurements of x-ray spectral line profiles [12]. However, the methods proposed in papers [8, 11] are easier to implement experimentally: it does not require performing the polarization measurements in the x-ray range, which would be relatively difficult to implement.

Thus, the novel methods presented in papers [8, 11] expand the possibilities for measuring super-strong magnetic fields up to ~10 GG expected to arise during relativistic laser–plasma interactions.

References

[1] Belyaev V S, Krainov V P, Lisitsa V S and Matafonov A P 2008 *Phys. Uspekhi* **51** 793
[2] Belyaev V S and Matafonov A P 2011 *Femtosecond-Scale Optics* ed A Andreev (Shanghai: InTech)

[3] Oks E, Dalimier E and Ya Faenov A *et al* 2017 *Opt. Express* **25** 1958
[4] Oks E, Dalimier E and Ya Faenov A *et al* 2017 *J. Phys. B: At. Mol. Opt. Phys.* **50** 245006
[5] Wagner U, Tatarakis M and Gopal A *et al* 2004 *Phys. Rev. E* **70** 026401
[6] Tatarakis M, Gopal A and Watts I *et al* 2002 *Phys. Plasmas* **9** 2244
[7] Tatarakis M, Watts I and Beg F N *et al* 2002 *Nature* **415** 280
[8] Dalimier E and Oks E 2018 *Atoms* **6** 60
[9] Gavrilenko V P 1988 *Sov. Phys. JETP* **67** 915
[10] Demkov Y, Monozon B and Ostrovsky V 1970 *Sov. Phys. JETP* **30** 775
[11] Oks E, Dalimier E and Angelo P 2019 *Spectrochim. Acta B* **157** 1
[12] Demura A V and Oks E 1998 *IEEE Trans. Plasma Sci.* **26** 1251

Chapter 6

Concluding remarks

This book is focused on *advances* in the x-ray spectroscopy of laser plasmas and is not intended to present the entire history of the x-ray spectroscopy of laser plasmas. It so happened that the advances were related mainly to the phenomena of the dips in x-ray spectral line profiles: mostly the Langmuir-wave-caused dips (L-dips), as presented in chapters 3–5, and also the charge-exchange caused dips (X-dips), as presented in chapter 2.

Here we list some works that did not fit into the format of chapters 2–5. First, we should mention paper by Gavrilenko, Faenov, Magunov *et al* [1]. In that work, the authors observed modulations in the O VIII Lyman-α line profiles emitted by clusters irradiated by femtosecond laser pulses. Specifically, they analyzed the experimental profiles of the Ly-alpha line of OVIII ions in N_2O clusters irradiated by femtosecond laser pulses of the intensity up to 4×10^{17} W cm^{-2}. The authors did not identify the L-dips of the X-dips. Instead they were talking about a sequence of local maxima and minima. By simulations the authors of paper [1] demonstrated that the allowance for some oscillating electric fields of the amplitude ∼(0.5–1.0) × 10^9 V cm^{-1} and of the frequency ∼(0.7–2.4) × 10^{15} s^{-1} leads to spectral features in the theoretical profiles of the Ly-alpha OVIII line that are similar to the features observed experimentally.

Another development concerns diagnostics of Langmuir solitons in laser plasmas. Langmuir solitons are relatively strong Langmuir waves in plasmas—the waves having a certain spatial formfactor (see, for example book [2]). In paper [3] the author calculated analytically the shape of satellites of dipole-forbidden lines in a spectrum *spatially-integrated* through a Langmuir soliton (or through a sequence of Langmuir solitons separated by some distance). The dipole-forbidden lines are the characteristic feature of He and Li spectral lines or of the spectral lines of He-like and Li-like ions. In paper [4] the author performed a general analysis of effects of Langmuir solitons on arbitrary spectral lines of hydrogen or hydrogenlike ions. He showed how the line profiles depend on the amplitude of the Langmuir solitons and

doi:10.1088/978-0-7503-3375-7ch6 © IOP Publishing Ltd 2020

on their separation from each other within the sequence of the solitons. These manifestations can be used for the experimental determination of the amplitude of the Langmuir solitons and of their separation from each other in their sequence. Further details are presented in appendix B of this book.

Yet another development relates to the situation where, at sufficiently high electron densities $N_e \gg 10^{18}$ cm^{-3}, mostly due to electron collisions the Langmuir turbulence can cease to be quasimonochromatic and rather represents a broadband electric field with the peak intensity at the plasma electron frequency $\omega_{pe} = (4\pi e^2 N_e / m_e)^{1/2}$, where e and m_e are the electron charge and mass, respectively. The electric field of the Langmuir turbulence can become multi-mode and can be even stochastic.

In this scenario the Langmuir turbulence can lead to a broadening of hydrogenic spectral lines. The first theoretical developments on this effect were presented in papers [5, 6] and then employed for analyzing some experimental results in papers [7–9]. A further theoretical study related to this physical situation was conducted in paper [10]. Specifically, the author of paper [10] considered modifications of profiles of hydrogen spectral lines under a multi-mode non-monochromatic linearly-polarized electric field. He obtained results for the case where the power spectrum of the stochastic electric field is Lorentzian. In paper [11] the author extended the results to the situation where the power spectrum of the stochastic electric field is Gaussian. Besides, he studied theoretically the general case of hydrogenlike spectral lines— rather than only hydrogen spectral lines considered in paper [10]. By analyzing the corresponding theoretical profiles of the hydrogenic Ly-beta line, the author of paper [11] proposed a new diagnostic method allowing for the first time not only to measure experimentally the average field of the Langmuir turbulence in dense plasmas, but also to find out the information on the power spectrum of the Langmuir turbulence. Further details are presented in appendix C of this book.

Last but not least: it is possible to perform experimentally the polarization analysis of x-ray spectral line profiles and in this way to increase the amount of the diagnostic information that can be deduced from laser–plasma experiments. The experimental demonstration of the principle of such a diagnostic was presented in papers [12, 13], where the experiments were performed at Z-pinches and the Ly-lines of Al XIII were observed. The experimental design of the x-ray polarization measurements was based on the following: if one observes a spectral line employing a crystal at the Bragg angle, and then observes the same line after rotating the crystal through 90°, the resulting spectra will correspond to two orthogonal linear polarizations. This can be achieved by using two spectrometers, positioned with respect to each other in such a way that the plasma line source is in the plane of the spectrometer in one case and perpendicular to it in the other case.

References

[1] Gavrilenko V P, Faenov A Y, Magunov A I, Pikuz T A, Skobelev I Y, Kim K Y and Milchberg H M 2006 *Phys. Rev.* A **73** 013203

[2] Kadomtsev B B 1982 *Collective Phenomena in Plasma* (Oxford: Pergamon)

[3] Oks E 2017 *J. Phys. Conf. Ser.* **810** 012006
[4] Oks E 2019 *Atoms* **7** 25
[5] Sholin G V 1970 *Sov. Phys. Doklady* **15** 1040
[6] Oks E and Sholin G V 1975 *Sov. Phys. JETP* **41** 482
[7] Zakatov L P, Plakhov A G, Shapkin V V and Sholin G V 1971 *Sov. Phys. Doklady* **16** 451
[8] Karfidov D M and Lukina N A 1997 *Phys. Lett.* A **232** 443
[9] Oks E 2016 *J. Phys. B: At. Mol. Opt. Phys.* **49** 065701
[10] Gavrilenko V P 1996 *Pis'ma v Zh. Tech. Phys. (Sov. Phys. Tech. Phys. Lett.)* **22** 23 (in Russian)
[11] Oks E 2020 *Spectrochim. Acta* B **167** 105815
[12] Clothiaux E J, Oks E, Weinheimer J, Svidzinski V and Schulz A 1997 *J. Quant. Spectrosc. Radiat. Transf.* **58** 531
[13] Weinheimer J, Oks E, Clothiaux E J, Schulz A and Svidzinski V 1998 *IEEE Trans. Plasma Sci.* **26** 1239

IOP Publishing

Advances in X-Ray Spectroscopy of Laser Plasmas

Eugene Oks

Appendix A

Overview of the theories of the dynamical Stark broadening of ion spectral lines in plasmas

Here we limit the presentation by the dynamical Stark broadening of hydrogenlike ion lines by plasma electrons. It was initially developed by Griem and Shen [1] (see also books [2, 3]). It is often called the conventional theory (CT); sometimes it is called the standard theory.

The dynamical broadening of spectral lines in plasmas by electrons is effective if the number ν_{We} of perturbing electrons in the sphere of the electron Weisskopf radius ρ_{We} is much smaller than unity (see, e.g. review [4]): $\nu_{We} = 4\pi N_e \rho_{We}^3/3 \ll 1$, where N_e is the electron density and $\rho_{We} \sim n^2 \, \hbar/(m_e v_{Te})$. Here n is the principal quantum number of the radiator energy level involved in the radiative transition and v_{Te} is the mean thermal velocity of plasma electrons. Under this condition, for the overwhelming majority of perturbing electrons, the characteristic frequency of the variation of the electron microfield $\Omega_e \sim v_{Te}/\rho_{We}$ is much greater than the instantaneous Stark splitting in the electron microfield. Physically, the electron Weisskopf radius is related to the impact parameters $\rho \sim \rho_{We}$ that contribute most effectively to the dynamical Stark broadening of spectral lines by electrons in plasmas [4].

The gist of dynamical effects in the Stark broadening of spectral lines in plasmas by electrons is the following. Collisions with plasma electrons cause *virtual transitions* mostly within the upper (n) and lower (n') multiplets during the radiative transition $n \leftrightarrow n'$. The primary outcome is a *decrease of the lifetime* of the states n' and/or n of the radiator, thus leading to the broadening of the corresponding spectral line.

The fact that virtual transitions occur mostly within the upper and lower multiplets conventionally leads to so-called *no-quenching approximation*, in which virtual transitions between states of different principal quantum numbers are totally disregarded. This approximation allows introducing the so-called *line space*: a direct product of the Hilbert space, spanned on the basis vectors of the n-shell, with the Hilbert space, spanned on the (complex-conjugated) basis vectors of the n'-shell.

doi:10.1088/978-0-7503-3375-7ch7   A-1   © IOP Publishing Ltd 2020

The origin of the CT can be traced back to the impact formalism developed by Baranger [5] and then by Kolb and Griem [6]. The central point of the impact formalism is the employment of a *coarse-grained time scale* Δt. Namely, on the one hand, Δt should be chosen such that it is much greater than the characteristic time ρ/v of the variation of the electric field created by the perturbing electron at the location of the radiating ion. Here ρ is the impact parameter and v is the velocity of the perturbing electron. On the other hand, Δt should be chosen such that it is much smaller than $[\max(\gamma, \Delta\omega, \omega_{\mathrm{pe}})]^{-1}$. Here γ is the inversed lifetime of the radiator (the impact width of the spectral line is of the order of γ), $\Delta\omega$ is the detuning from the unperturbed frequency ω_0 of the spectral line, $\omega_{\mathrm{pe}} = (4\pi N_e e^2/m_e)^{1/2}$ is the plasma electron frequency.

Physically, the coarse-grained time scale means that we are not interested in details of the evolution of the radiator during the characteristic time of the individual collision ρ/v. Instead, we are interested in the evolution of the radiator during larger time intervals Δt. The limits of validity of the impact approximation are controlled by the introduction of the coarse-grained time.

For completeness we should mention the so-called unified formalism developed by Vidal, Cooper, and Smith [7]. The primary distinction of the unified formalism from the impact formalism is the following. The *impact* formalism considers only *completed collisions*, while the *unified* formalism allows also for *incomplete collisions*. Another difference relates to the fact that the unified formalism allows (at least in principle) a transition to the nearest-neighbor quasistatic result in the wings of the spectral line. This difference is less important: quantitatively the unified formalism does not always produce such transition correctly. It should be noted that one of the conditions for introducing the coarse-grain time scale is somewhat relaxed in the unified formalism compared to the impact formalism. Namely, in the unified formalism it is required that $\rho/v \leqslant \Delta t$—compared to the requirement $\rho/v \ll \Delta t$ in the impact formalism. Further details on the rigorous description of both formalisms can be found in the comprehensive review by Sahal-Brechot [8].

The CT leads to the expression for the so-called electron impact broadening operator containing a diverging integral: the integral over the impact parameters. This integral diverges at both small and large impact parameters. This divergence is one of the primary deficiencies of the CT.

The divergence at large ρ is related to the long-range nature of the Coulomb potential. There occurs the plasma screening of the electric field of the perturbing electron at the distances larger than the Debye radius

$$\rho_D = [T_e/(^4pe^2 N_e)]^{1/2} \tag{A.1}$$

where T_e is the electron temperature. This leads to choosing the upper cutoff at $\rho_{\max} \sim \rho_D$. In distinction, the divergence at small impact parameters in the CT is due the employment of the perturbation expansion.

The above deficiency of the CT has been eliminated with the development of the so-called generalized theory (GT) of the dynamical Stark broadening of hydrogenic spectral lines [9, 10] (see also book [11]). The GT is based on a *generalization* of the formalism of dressed atomic states (DAS) in plasmas.

Originally the DAS-formalism was introduced for studies of the interaction of a *monochromatic* (or quasi-monochromatic) field—such as, for example, laser or maser radiation—with gases. In the course of further research, the DAS-formalism was extended to the description of the interaction of a laser or maser radiation with plasmas [12]. This resulted in more accurate analytical calculations and better (more robust) codes.

The *generalization* of DAS in papers [9, 10] was based on utilizing atomic states dressed not by a monochromatic field, but by a *broad-band* field of plasma electrons and ions. As should be expected, the generalized DAS is a more complex concept than the usual DAS.

Due to the utilization of the generalized DAS, the authors of papers [9, 10] succeeded in taking into account analytically a *coupling* of the electron and ion microfields facilitated by the radiating atom. This coupling increases with the growth of the principal quantum number n and with the growth of the electron density N_e, as well as with the decrease of the temperature T.

We mention in passing that some of the later extensions of the GT caused some discussions in the literature. However, the core GT developed in papers [9, 10] has stood the test of time.

The GT eliminated large discrepancies—up to a factor of two—between the CT and benchmark experiments, as presented in part of book [11]. Below are some more details on the GT.

The Hamiltonian of a hydrogenic atom or ion subjected to the quasistatic part $\mathbf{F}$ of the ion microfield and to the electron-produced dynamic field $\mathbf{E}(t)$ can be represented in the form

$$H = H_0 - \mathbf{dF} - \mathbf{dE}(t), \tag{A.2}$$

where H_0 is the unperturbed Hamiltonian, $\mathbf{d}$ is the dipole moment operator. One chooses the axis Oz of the parabolic quantization along the field $\mathbf{F}$. Then the operator $-\mathbf{dF}$ is diagonal in any subspace of a fixed principal quantum number n. For this reason, in the CT this interaction was taken into account 'exactly' (neglecting only some corrections due to the matrix elements of the operator d_z corresponding to $\Delta n \neq 0$). The interaction with the field $\mathbf{E}(t)$ in the CT was subsequently treated in the second order of the Dirac's (time-dependent) perturbation theory.

In distinction, in the GT, the entire z-component of the total field $\mathbf{F} + \mathbf{E}(t)$ is allowed for in the same (or analogous) way as the field $\mathbf{F}$ was treated in the CT. This was possible because the interaction $-d_z[F + E_z(t)]$ (and not only its part $-d_z F$) is diagonal in any n-subspace. Therefore, the z-component of the electron microfield can be taken into account much more accurately than in the CT. In this way, the Stark sublevels are *dynamically dressed* by the entire z-component of the total microfield $\mathbf{F} + \mathbf{E}(t)$. This is the central feature of the GT that leads to its several advantages over the CT.

In the GT electron broadening operator and in the GT correlation function, whose Fourier transform is the lineshape, there are adiabatic and nonadiabatic terms. The adiabatic term in the correlation function is proportional to the part $d_z d_z$ of the operator **dd**; the operator $d_z d_z$ is diagonal in the line space (we recall that the line space is the Gilbert space spanned on the upper and lower Stark substates involved in the radiative transition). The rest of the correlation function corresponds to the non-adiabatic contribution: it is proportional to the operator $d_x d_x + d_y d_y$ that has both diagonal and non-diagonal matrix elements in the line space.

In the GT, the adiabatic part of the electron broadening operator and of the correlation function is calculated *exactly*. This exact result is equivalent to the summation of all orders of the corresponding Dyson expansion entering formulas of both the CT and the GT. This is one of the most important distinctions from the CT, where all terms in the electron broadening operator and of the correlation function are calculated only in the first non-vanishing (namely, the second) order of the Dyson expansion.

As a result, the GT is *convergent* at small impact parameters while the corresponding CT for neutral radiators is divergent. (For charged radiators the CT is formally convergent, but leads quantitatively to wrong results.) It is the allowance for the 'dressing' by the broad band field $F + E_z(t)$ that eliminates the divergence and enhances the accuracy of the results. In distinction, the higher the electron density and/or the principal quantum number (or the lower the temperature), the greater becomes the inaccuracy of the CT.

It should be noted that the GT was developed *analytically* to the same level as the CT. This is counterintuitive because the starting formulas for the GT are more complicated than for the CT.

Speaking specifically about the dynamical Stark broadening of hydrogenlike spectral lines by plasma electrons, there is another fundamental flaw of the CT. This flaw has been eliminated in paper [13] (see also chapter 12 of book [14]), which we follow below.

In the CT, the perturbing electrons are considered moving along hyperbolic trajectories in the Coulomb field of the effective charge $Z - 1$ (in atomic units), where Z is the nuclear charge of the radiating ion. In other words, in the CT there was made a simplifying assumption that the motion of the perturbing electron can be described in frames of a two-body problem, one particle being the perturbing electron and the other 'particle' being the charge $Z - 1$.

However, in reality one has to deal with a three-body problem: the perturbing electron, the nucleus, and the bound electron. Therefore, trajectories of the perturbing electrons should be more complicated.

In paper [13] the authors took this into account by using the standard analytical method of separating rapid and slow subsystems—see, e.g. book [15]. The characteristic frequency of the motion of the bound electron around the nucleus is much higher than the characteristic frequency of the motion of the perturbing electron around the radiating ion. Therefore, the former represents the rapid subsystem and the latter represents the slow subsystem. This approximate analytical method allows a sufficiently accurate treatment in situations where the perturbation theory fails—see, e.g. book [15].

By applying this method the authors of paper [13] obtained more accurate analytical results for the electron broadening operator than in the CT. They showed by examples of the electron broadening of the Lyman lines of He II that the allowance for this effect increases with the electron density N_e, becomes significant already at $N_e \sim 10^{17}$ cm^{-3} and very significant at higher densities. Below are the details.

In the CT the electron broadening operator is expressed in the form (see, e.g. paper [1])

$$\Phi_{ab} \equiv 2\pi v N_e \int d\rho \, \rho \{ S_a S_b^* - 1 \}, \tag{A.3}$$

where N_e, v, and ρ are the electron density, velocity, and impact parameter, respectively; $S_a(0)$ and $S_b(0)$ are the S matrices for the upper (a) and lower (b) states involved in the radiative transition, respectively; $\{\dots\}$ stands for the averaging over angular variables of vectors $\mathbf{v}$ and $\boldsymbol{\rho}$. Further in the CT, the collisions are subdivided into weak and strong. The weak collisions are treated by the time-dependent perturbation theory. The impact parameter, at which the formally calculated expression $\{S_a S_b^* - 1\}$ for a weak collision starts violating the unitarity of the S-matrices, serves as the boundary between the weak and strong collisions and is called Weisskopf radius ρ_{We}.

So, in the CT the integral over the impact parameter diverges at small ρ. Therefore, in the CT this integral is broken down into two parts: from 0 to ρ_{We} (strong collisions) and from ρ_{We} to $\rho_{\max}$ for weak collisions. The upper cutoff $\rho_{\max}$ (typically chosen to be the Debye radius given by equation (A.1)) is necessary because this integral diverges also at large ρ.

In the CT, after calculating the S matrices for weak collisions, the electron broadening operator becomes (*in atomic units*)

$$\Phi_{ab}{}^{weak} \equiv C \int_{\rho_{we}}^{\rho_{\max}} d\rho \, \rho \sin^2 \frac{\Theta(\rho)}{2} = \frac{C}{2} \int_{\Theta_{\min}}^{\Theta_{\max}} d\Theta \frac{d\rho^2}{d\Theta} \sin^2 \frac{\Theta}{2}, \tag{A.4}$$

where Θ is the scattering angle for the collision between the perturbing electron and the radiating ion (the dependence between Θ and ρ being discussed below) and the plasma electron and the operator C is

$$C = -\frac{4\pi}{3} N_e \left[\int_0^\infty dv v^3 f(v) \right] \frac{m^2}{(Z-1)^2} (r_a - r_b^*)^2. \tag{A.5}$$

Here $f(v)$ is the velocity distribution of the perturbing electrons, r is the radius-vector operator of the bound electron (which scales with Z as $1/Z$), and m is the reduced mass of the system 'perturbing electron—radiating ion'.

In the CT the scattering occurs in the effective Coulomb potential, so that the trajectory of the perturbing electron is hyperbolic and the relation between the impact parameter and the scattering angle is given by

$$\rho^{(0)} = \frac{Z-1}{mv^2}\cot\frac{\Theta}{2}.$$

(A.6)

In paper [13] the authors considered the realistic situation where trajectories of the perturbing electrons are more complicated because the perturbing electron, the nucleus, and the bound electron should be more accurately treated as the three-body problem. We use the standard analytical method of separating rapid and slow subsystems—see, e.g. book [15]. It is applicable here because the characteristic frequency v_{Te}/ρ_{We} of the variation the electric field of the perturbing electrons at the location of the radiating ion is much smaller than the frequency Ω_{ab} of the spectral line (the latter, e.g. in case of the radiative transition between the Rydberg states would be the Keppler frequency or its harmonics)—more details on this are presented at the end of this chapter.

The first step in this method is to 'freeze' the slow subsystem (perturbing electron) and to find the analytical solution for the energy of the rapid subsystem (the radiating ion) that would depend on the frozen coordinates of the slow subsystem (in our case it will be the dependence on the distance R of the perturbing electron from the radiating ion). To the first non-vanishing order of the R-dependence, the corresponding energy in the parabolic quantization is given by

$$E_{nq}(R) = -\frac{Z^2}{n^2} + \frac{3\,nq}{2\,ZR^2},$$

(A.7)

where n and $q = n_1 - n_2$ are the principal and electric quantum numbers, respectively; n_1 and n_2 are the parabolic quantum numbers.

The next step in this method is to consider the motion of the slow subsystem (perturbing electron) in the 'effective potential' $v_{\text{eff}}(R)$ consisting of the actual potential plus $E_{nq}(R)$. Since the constant term in equation (A.7) does not affect the motion, the effective potential for the motion of the perturbing electron can be represented in the form

$$V_{\text{eff}}(R) = -\frac{\alpha}{R} + \frac{\beta}{R^2},\quad \alpha = Z - 1.$$

(A.8)

For the spectral lines of the Lyman series, since the lower (ground) state b of the radiating ion remains unperturbed (up to/including the order $\sim 1/R^2$), the coefficient β is

$$\beta = \frac{3\,n_a q_a}{2\,Z}.$$

(A.9)

For other hydrogenic spectral lines, for taking into account both the upper and lower states of the radiating ion, the coefficient β can be expressed as

$$\beta = \frac{3\,(n_a q_a - n_b q_b)}{2\,Z}.$$

(A.10)

The motion in the potential from equation (A.8) allows an exact analytical solution. In particular, the relation between the scattering angle and the impact parameter is no longer given by equation (A.6), but rather becomes (see, e.g. book [16])

$$\Theta = \pi - \frac{2}{\sqrt{1 + \frac{2\,m\beta}{M^2}}} \arctan \sqrt{\frac{4\,E}{\alpha^2}\left(\beta + \frac{M^2}{2m}\right)}. \tag{A.11}$$

Here E and M are the energy and the angular momentum of the perturbing electron, respectively. One can rewrite the angular momentum in terms of the impact parameter ρ as

$$M = mv\rho \tag{A.12}$$

Then a slight rearrangement of equation (A.11) yields

$$\tan\left(\frac{\pi - \Theta}{2}\sqrt{1 + \frac{2\,\beta}{mv^2\rho^2}}\right) = \frac{v}{\alpha}\sqrt{m^2v^2\rho^2 + 2\,m\beta}. \tag{A.13}$$

After solving equation (A.13) for ρ and substituting the outcome in equation (A.4), a more accurate expression for the electron broadening operator can be obtained. However, equation (A.13) does not have an exact analytic solution for ρ so that this could be done only numerically.

In paper [13], for getting the message across in the simplest form, the authors provided an approximate analytical solution of equation (A.13) by expanding it in powers of β. This yields (keeping up to the first power of β)

$$\tan\left(\frac{\pi - \Theta}{2}\right) + \left(\frac{\pi - \Theta}{2}\right)\left[1 + \tan^2\left(\frac{\pi - \Theta}{2}\right)\right]\frac{\beta}{mv^2\rho^2} \approx \frac{mv^2\rho}{\alpha} + \frac{\beta}{\alpha\rho}. \tag{A.14}$$

The authors of paper [13] were seeking the analytical solution for ρ in the form $\rho \approx \rho^{(0)} + \rho^{(1)}$, where $\rho^{(0)}$ corresponds to $\beta = 0$ (and was given by equation (A.6)) and $\rho^{(1)} \ll \rho^{(0)}$. Substitution of $\rho \approx \rho^{(0)} + \rho^{(1)}$ into equation (A.14) yields the expression

$$\frac{(\pi - \Theta)\,\beta}{2\,mv^2\rho^{(0)2}\sin^2\frac{\Theta}{2}} - \frac{\beta}{\alpha\rho^{(0)}} \approx \frac{mv^2\rho^{(1)}}{\alpha}. \tag{A.15}$$

After solving equation (A.15) for $\rho^{(1)}$, one gets the expression for ρ:

$$\rho \approx \frac{\alpha}{mv^2}\cot\frac{\Theta}{2} + \frac{\beta}{\alpha}\left(\frac{\pi - \Theta}{2\cos^2\frac{\Theta}{2}} - \tan\frac{\Theta}{2}\right). \tag{A.16}$$

As a reminder, the goal is to perform the integration in equation (A.3) for obtaining a more accurate analytical result for the electron broadening operator. This can be more easily accomplished by performing the integration over Θ instead of ρ. For this purpose, first, the authors of paper [13] squared equation (A.16)

$$\rho^2 \approx \frac{\alpha^2}{m^2 v^4}\cot^2\frac{\Theta}{2} + \frac{\beta}{mv^2}\left(\frac{\pi - \Theta}{\sin\frac{\Theta}{2}\cos\frac{\Theta}{2}} - 1\right), \tag{A.17}$$

where only the first order terms in β have been kept for consistency. To make formulas simpler, they denoted $\phi = \Theta/2$. After differentiating equation (A.17) with respect to ϕ, the authors of paper [13] obtained

$$\frac{d\rho^2}{d\phi} \approx -\frac{\alpha^2}{m^2 v^4}\frac{2\cot\phi}{\sin^2\phi} - \frac{2\beta}{mv^2}\left[\left(\frac{1}{\sin\phi\cos\phi}\right) + \left(\frac{\pi}{2} - \phi\right)\left(\frac{1}{\sin^2\phi} - \frac{1}{\cos^2\phi}\right)\right]. \tag{A.18}$$

After substituting in the utmost right side of equation (A.4) first $\Theta = 2\phi$ and then $\frac{d\rho^2}{d\phi}$ from equation (A.18), the contribution of the weak collisions to the electron broadening operator becomes

$$\Phi_{ab}{}^{\text{weak}} = -C\left[\frac{\alpha^2}{m^2 v^4}\int_{\phi_{\min}}^{\phi_{\max}}\cot\phi\, d\phi + \frac{\beta}{mv^2}\int_0^{\frac{\pi}{2}}\tan\phi\, d\phi \right.$$
$$\left. + \frac{\beta}{mv^2}\int_0^{\frac{\pi}{2}}\left(\frac{\pi}{2} - \phi\right)(1 - \tan^2\phi)d\phi\right]. \tag{A.19}$$

In equation (A.19), in the two correction terms proportional to β, The authors of paper [13] extended the integration over the full range of the variation of the angle ϕ. The corresponding minor inaccuracy would not contribute significantly to the electron broadening operator, since the terms involving β are considered to be a relatively small correction to the first term in equation (A.19).

Performing the integrations in equation (A.19), they obtained:

$$\Phi_{ab}{}^{\text{weak}} = -\frac{4\pi}{3}N_e(\mathbf{r}_a - \mathbf{r}_b^*)^2\left[\int_0^\infty dv\frac{f(v)}{v}\right]$$
$$\times\left[\log\frac{\sin\phi_{\max}}{\sin\phi_{\min}} + \frac{mv^2\beta}{(Z-1)^2}\left(\frac{\pi^2}{4} - 1\right)\right]. \tag{A.20}$$

Here and below the expression $(\mathbf{r}_a - \mathbf{r}_b^*)^2$ stands for the scalar product (also known as the dot-product) of the operator $(\mathbf{r}_a - \mathbf{r}_b^*)$ with itself. In the theory of the dynamical Stark broadening of spectral lines in plasmas by electrons, the corresponding matrix elements are calculated with respect to the unperturbed wave functions.

Then the authors of paper [13] added the CT estimate for the contribution of strong collisions

$$\Phi_{ab}{}^{\text{strong}} \approx \pi v N_e \rho_{We}{}^2, \tag{A.21}$$

where ρ_{W_e} corresponds to $\phi_{\max}$. Expressions for $\phi_{\max}$ and $\phi_{\min}$ are given in paper [1] (in equations (9) and (10a)) as follows

$$\sin\phi_{\max} = \sqrt{\frac{3}{2}}\,\frac{Z(Z-1)}{(n_a^2 - n_b^2)mv},\tag{A.22}$$

$$\sin\phi_{\min} = \frac{\dfrac{Z-1}{mv^2\rho_D}}{\sqrt{1 + \dfrac{(Z-1)^2}{m^2v^4\rho_D^2}}}\tag{A.23}$$

It should be emphasized that the factor $(n_a^2 - n_b^2)$ in the denominator of the right side of equation (A.22) was an approximate allowance by the authors of paper [1] for the contribution of the lower level b while estimating the operator $(r_a - r_b^*)$ for hydrogenic lines of spectral series other than the Lyman lines. However, for the Lyman lines the lower (ground) level does not contribute to electron broadening operator, so that for the Lyman lines equation (A.22) should be simplified as follows:

$$\sin\phi_{\max} = \sqrt{\frac{3}{2}}\,\frac{Z(Z-1)}{n_a^2 mv}.\tag{A.24}$$

It should be noted that at relatively small velocities of perturbing electrons, the right side of equation (A.22) or equation (A.24) could exceed unity. In this case one should set $\sin\phi_{\max} = 1$, what corresponds to $\rho_{\min} = 0$, so that there would be no contribution from strong collisions. Typically, the range of such small velocities has a very low statistical weight in the electron velocity distribution.

After substituting the above formulas for $\sin\phi_{\max}$ and $\sin\phi_{\min}$ into equation (A.19), and combining the contributions from weak and strong collisions, the authors of paper [13] obtained the final results for the electron broadening operator:

$$\begin{aligned}
\Phi_{ab}(\beta) = -\frac{4\pi}{3}N_e(r_a - r_b^*)^2\Bigg[\int_0^\infty dv\frac{f(v)}{v}\Bigg]\Bigg\{\frac{1}{2}\Bigg[1 - \frac{3}{2}\frac{Z^2(Z-1)^2}{(n_a^2 - n_b^2)^2 m^2 v^2}\Bigg] \\
+ \log\Bigg[\sqrt{\frac{3}{2}}\frac{Zv\rho_D}{(n_a^2 - n_b^2)}\sqrt{1 + \Big(\frac{Z-1}{mv^2\rho_D}\Big)^2}\Bigg] + \frac{mv^2\beta}{(Z-1)^2}\Big(\frac{\pi^2}{4} - 1\Big)\Bigg\}
\end{aligned}\tag{A.25}$$

for the non-Lyman lines and

$$\begin{aligned}
\Phi_{ab}(\beta) = -\frac{4\pi}{3}N_e(r_a - r_b^*)^2\Bigg[\int_0^\infty dv\frac{f(v)}{v}\Bigg]\Bigg\{\frac{1}{2}\Bigg[1 - \frac{3}{2}\frac{Z^2(Z-1)^2}{n_a^4 m^2 v^2}\Bigg] \\
+ \log\Bigg[\sqrt{\frac{3}{2}}\frac{Zv\rho_D}{n_a^2}\sqrt{1 + \Big(\frac{Z-1}{mv^2\rho_D}\Big)^2}\Bigg] + \frac{mv^2\beta}{(Z-1)^2}\Big(\frac{\pi^2}{4} - 1\Big)\Bigg\}
\end{aligned}\tag{A.26}$$

for the Lyman lines. Here and below $\log[\ldots]$ stands for the natural logarithm.

In order to determine the significance of this effect, it is necessary then to evaluate the ratio

$$\text{ratio} = \frac{\dfrac{3}{2}\dfrac{mv^2(n_a q_a - n_b q_b)}{(Z-1)^2}\left(\dfrac{\pi^2}{4}-1\right)}{\dfrac{1}{2}\left[1 - \dfrac{3}{2}\dfrac{Z^2(Z-1)^2}{\left(n_a^2 - n_b^2\right)^2 m^2 v^2}\right] + \log\left[\sqrt{\dfrac{3}{2}}\dfrac{Zv\rho_D}{\left(n_a^2 - n_b^2\right)}\sqrt{1 + \left(\dfrac{Z-1}{mv^2\rho_D}\right)^2}\right]} \quad (A.27)$$

for the non-Lyman lines or the ratio

$$\text{ratio} = \frac{\dfrac{3}{2}\dfrac{mv^2 n_a q_a}{(Z-1)^2}\left(\dfrac{\pi^2}{4}-1\right)}{\dfrac{1}{2}\left[1 - \dfrac{3}{2}\dfrac{Z^2(Z-1)^2}{n_a^4 m^2 v^2}\right] + \log\left[\sqrt{\dfrac{3}{2}}\dfrac{Zv\rho_D}{n_a^2}\sqrt{1 + \left(\dfrac{Z-1}{mv^2\rho_D}\right)^2}\right]} \quad (A.28)$$

for the Lyman lines.

Below we reproduce numerical examples for several Lyman lines from paper [13]. As is customary in the Stark broadening theory, instead of the integration over velocities, for the numerical examples the authors of paper [13] used the mean thermal velocity v_T of the perturbing electrons. In atomic units, the mean thermal velocity v_T, the Debye radius ρ_D, and the reduced mass can be expressed as follows

$$v_T = 0.1917\sqrt{\frac{T(\text{eV})}{m}} \qquad \rho_D = 1.404 \times 10^{11}\sqrt{\frac{T(\text{eV})}{N_e(\text{cm}^{-3})}} \qquad m = \frac{1 + \dfrac{m_e}{Am_p}}{1 + \dfrac{2\,m_e}{Am_p}}, \quad (A.29)$$

where m_e is the electron mass, m_p is the proton mass, and A is the atomic number of the radiating ion ($A \approx 2Z$).

Table A1 presents the values of the ratio from equation (A.28) for several Lyman lines of He II at the temperature $T = 8$ eV and the electron density $N_e = 2 \times 10^{17}$ cm^{-3}.

Figure A1 shows the ratio from equation (A.28) versus the electron density N_e for the Stark components of the electric quantum number $|q| = 1$ of Lyman-alpha ($n = 2$), Lyman-beta ($n = 3$), and Lyman-gamma ($n = 4$) lines of He II at the temperature $T = 8$ eV.

It is seen that for the electron broadening of the Lyman lines of He II, the allowance for the effect under consideration indeed becomes significant already at electron densities $N_e \sim 10^{17}$ cm^{-3} and increases with the growth of the electron density. It should be noted that when the ratio, formally calculated by equation (A.28), becomes comparable to unity, this is the indication that the approximate analytical treatment based on expanding equation (A.13) up to the first order of parameter β, is no longer valid. In this case the calculations should be based on solving equation (A.13) with respect to ρ without such approximation.

Table A1. Ratio from equation (A.28) for the Stark components of several Lyman lines of He II. at the temperature $T = 8$ eV and the electron density $N_e = 2 \times 10^{17}$ cm^{-3} [13].

n	$|q|$	$ratio$
2	1	0.3261
3	1	0.3748
3	2	0.7496
4	1	0.5156
4	2	1.0311
4	3	1.5467

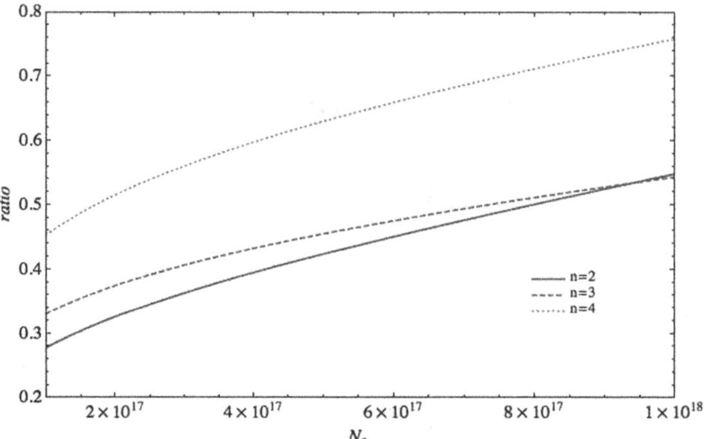

Figure A1. Ratio from equation (A.28) versus the electron density N_e for the Stark components of the electric quantum number $|q| = 1$ of Lyman-alpha ($n = 2$), Lyman-beta ($n = 3$), and Lyman-gamma ($n = 4$) lines of He II at the temperature $T = 8$ eV [13].

Thus, the authors of paper [13] obtained more accurate analytical results for the electron broadening operator compared to the CT. By examples of the electron broadening of the Lyman lines of He II, we demonstrated that the allowance for this effect becomes significant at electron densities $N_e \sim 10^{17}$ cm^{-3} and very significant at higher densities. It is well-known that for relatively low-Z radiators, the broadening by electrons is comparable to the broadening by ions, so that the correction to the broadening by electrons, introduced in the present paper, should be significant for the total Stark width.

It is important to emphasize that the authors of paper [13] were able to obtain the above analytical results primarily due to the underlying fundamental symmetry of the class of potentials $v(R) = -A/R + B/R^2$, where A and B are constants. Namely, this class of potentials possesses an additional conserved quantity $M_{\text{eff}}^2 = M^2 +$

$2mB$, where M is the angular momentum and m is the mass of a particle, so that M_{eff} is the effective angular momentum. As for the impact approximation, it was not crucial to work [13]—the authors used it only for the following two purposes: first, to get the message across in a simple form, and second, for the comparison with the CT (in which the impact approximation was crucial), so that we would compare 'apples to apples' rather than 'apples to oranges'.

The authors of paper [13] also mentioned that in 1981, Baryshnikov and Lisitsa [17] published very interesting results for the electron broadening of hydrogen-like spectral lines in plasmas (also presented later in book [18]) in frames of the quantum theory of the dynamical Stark broadening, while we obtained our results in frames of the semiclassical theory of the dynamical Stark broadening, just as in the CT. (For clarity: in the semiclassical theory, the radiating atom/ion is treated quantally, and perturbing electrons classically; in the quantum theory both the radiating atom/ion and perturbing electrons are treated quantally.) Both in paper [17] and in paper [13], there was used the underlying symmetry of the class of potentials $v(R) = -A/R + B/R^2$ for obtaining analytical solutions.

A specific result for the line width Baryshnikov and Lisitsa [17] obtained for Lyman lines in the classical limit using the impact approximation, as presented in their equations (4.5) and (4.6). The authors of paper [13] compared their results from equations (4.5) and (4.6) with the CT [1] for He II Lyman lines. It turned out that for $N_e \sim (10^{17}\text{--}10^{18})$ cm^{-3}, i.e. for the range of electron densities, in which the overwhelming majority of measurements of the width of He II lines were performed, Baryshnikov–Lisitsa's line width exceeds the CT line width by two orders of magnitude or more. In view of the fact that the width of He II lines, measured by various authors in benchmark experiments (i.e. experiments where plasma parameters were measured independently of the line widths), never exceeded the CT width by more than a factor of two (see, e.g. benchmark experiments [19–21]), this seems to indicate that something might be incorrect in equations (4.5) and (4.6) from paper [17] (though methodologically it was a very interesting paper). In distinction, the corrections to the CT introduced in paper [13], do not exceed the factor of two for He II lines in the range of $N_e \sim (10^{17}\text{--}10^{18})$ cm^{-3}.

Finally, the authors of paper [13] provided a detailed proof of the applicability of the analytical method (that they used) as follows. The characteristic frequency of the motion of the perturbing electron around the radiating ion in the process of the Stark broadening of spectral lines is the so-called Weisskopf frequency

$$\omega_{We} = \frac{v_T}{\rho_{We}} \sim \frac{Zmv_T^2}{(n_a^2 - n_b^2)\hbar} \sim \frac{ZT}{(n_a^2 - n_b^2)\hbar}. \tag{A.30}$$

The characteristic frequency of the motion of the bound electron around the nucleus is the frequency of the spectral line

$$\Omega = \frac{Z^2 U_H}{\hbar}\left(\frac{1}{n_b^2} - \frac{1}{n_a^2}\right), \tag{A.31}$$

where U_H is the ionization potential of hydrogen. The ratio of these two frequencies is

$$\frac{\omega_{We}}{\Omega} \sim \left(\frac{T}{ZU_H}\right)\left[\frac{n_a^2 n_b^2}{(n_a^2 - n_b^2)^2}\right]. \tag{A.32}$$

For the simplicity of estimating this ratio, the authors of paper [13] considered $n_a \gg n_b$, so that

$$\frac{\omega_{We}}{\Omega} \sim \left(\frac{T}{Zn_a^2 U_H}\right) \ll 1 \tag{A.33}$$

as long as

$$T(\text{eV}) \ll (13.6\text{eV})Zn_a^2. \tag{A.34}$$

For example, for $Z = 2$ the above validity condition becomes

$$T(\text{eV}) \ll (27.2\text{eV})n_a^2 \tag{A.35}$$

and is satisfied for a broad range of temperatures, at which He II spectral lines are observed in plasmas.

References

[1] Griem H R and Shen K Y 1961 *Phys. Rev.* **122** 1490

[2] Griem H R 1964 *Plasma Spectroscopy* (New York: McGraw-Hill)

[3] Griem H R 1974 *Spectral Line Broadening by Plasmas* (Cambridge, MA: Academic)

[4] Lisitsa V S 1977 *Sov. Phys. Uspekhi* **122** 603

[5] Baranger M 1958 *Phys. Rev.* **111** 481 494
    Baranger M 1958 *Phys. Rev.* **112** 855

[6] Kolb A C and Griem H R 1958 *Phys. Rev.* **111** 514

[7] Vidal C R, Cooper J and Smith E W 1970 *J. Quant. Spectrosc. Radiat. Transf.* **10** 197111 263

[8] Sahal-Brechot S 1969 *Astron. Astrophys.* **1** 91

[9] Ispolatov Y and Oks E 1994 *J. Quant. Specrosc. Radiat. Transf.* **51** 129

[10] Oks E, Derevianko A and Ispolatov Y 1995 *J. Quant. Specrosc. Radiat. Transf.* **54** 307

[11] Oks E 2006 *Stark Broadening of Hydrogen and Hydrogenlike Spectral Lines in Plasmas: The Physical Insight* (Oxford: Alpha Science International)

[12] Oks E 1995 *Plasma Spectroscopy: the Influence of Microwave and Laser Fields* (Berlin: Springer)

[13] Sanders P and Oks E 2018 *J. Phys. Commun.* **2** 035033

[14] Oks E 2019 *Analytical Advances in Quantum and Celestial Mechanics: Separating Rapid and Slow Subsystems* (Bristol: IOP Publishing)

[15] Galitski V, Karnakov B, Kogan V and Galitski V Jr 2013 *Exploring Quantum Mechanics* (Oxford: Oxford University Press) problem 8.55.

[16] Kotkin G L and Serbo V G 1971 *Collection of Problems in Classical Mechanics* (Oxford: Pergamon) problem 2.3.

[17] Baryshnikov F F and Lisitsa V S 1981 *Sov. Phys. JETP* **53** 471

[18] Bureyeva L A and Lisitsa V S 2000 *A Perturbed Atom, Astrophysics and Space Physics Reviews* (Boca Raton, FL: CRC Press)
[19] Grützmacher K and Johannsen U 1993 *Spectral Line Shapes* vol 7, ed R Stamm and B Talin (New York: Nova Science) 139
[20] Ahmad R 1999 *Eur. Phys. J.* D **7** 123
[21] Wrubel T, Büscher S, Kunze H-J and Ferri S 2001 *J. Phys.* B **34** 461

Appendix B

Diagnostic of Langmuir solitons in plasmas by using hydrogenic spectral lines

Langmuir solitons are relatively strong Langmuir waves in plasmas—the waves having a certain spatial formfactor (see, for example book [1]). There are only very few theoretical papers on their spectroscopic diagnostics in plasmas—in distinction to a large number of works on various spectroscopic diagnostics of relatively weak Langmuir waves in plasmas and their successful implementation (see, e.g. books [2–4] and references therein).

In paper [5] the author calculated analytically the shape of satellites of dipole-forbidden lines in a spectrum *spatially-integrated* through a Langmuir soliton (or through a sequence of Langmuir solitons separated by a distance L). The dipole-forbidden lines are the characteristic feature of He and Li spectral lines or of the spectral lines of He-like and Li-like ions. In the profiles of these spectral lines, Langmuir waves can give rise to satellites of the dipole-forbidden components of these lines. The primary outcome of paper [5] (presented also in section 7.3 of book [2]) was the following.

In the case of Langmuir solitons, the peak intensity of the satellites of the dipole-forbidden lines can be significantly enhanced—by orders of magnitude—compared to the case of non-solitonic Langmuir waves. This specific feature allows distinguishing Langmuir solitons from non-solitonic Langmuir waves.

In distinction to the above method based on the dipole-forbidden spectral lines of He, Li, as well as He-like and Li-like ions, speaking of using hydrogenic spectral lines, one should mention that Hannachi *et al* [6] performed simulations for finding the effect of Langmuir solitons on the hydrogen Ly_α line. The outcome was an additional broadening. However, even at the low electron density $N_e = 10^{14}$ cm^{-3}, this additional broadening was very small compared to the Stark broadening by plasma microfields. Moreover, the additional broadening rapidly diminished with the increase of N_e—as a result, there would be practically no additional broadening at $N_e > 10^{15}$ cm^{-3}. In paper [7] Hannachi *et al* introduced into consideration additionally a magnetic field. Hannachi *et al* [7] performed simulations at the electron density $N_e = 10^{13}$ cm^{-3}. However, this electron density is unrealistically low for the modern tokamaks (applications to which were hoped for by the authors of paper [7]) and again, the additional broadening would rapidly diminish for more realistic values of the electron density (that is, for higher values of the electron density). Therefore, it appears highly questionable that the results by Hannachi *et al* [6, 7] could be of any use for the experimental diagnostics of Langmuir solitons. The results from paper [6] were also reproduced in one part of paper [8] by Stamm *et al*[1].

In paper [10], which we follow here, the author conducted a general study effects of Langmuir solitons on arbitrary spectral lines of hydrogen or hydrogen-like ions. Then by using the Ly-beta line as an example, he compared the main features of the profiles for the case of the Langmuir solitons with the case of the non-solitonic Langmuir waves of the same amplitude. He demonstrated how the line profiles depend on the amplitude of the Langmuir solitons and on their separation from each other within the sequence of the solitons.

Langmuir solitons (or a set of Langmuir solitons separated by a distance L) have the following spatial formfactor [1]:

$$F(x, t) = E(x) \cos \omega t, \quad E(x) = E_0/ch(x/\lambda), \quad \lambda \ll L. \tag{B.1}$$

Here

$$\omega = \omega_{\mathrm{pe}} - 3Te/\left(2m_e\omega_{\mathrm{pe}}\lambda^2\right), \tag{B.2}$$

where

$$\omega_{\mathrm{pe}} = (4\pi e^2 N_e/m_e)^{1/2} \tag{B.3}$$

is the plasma electron frequency (m_e being the electron mass) and λ is the characteristic size of the soliton. A diagnostic of Langmuir solitons consists not only in finding experimentally the electric field oscillating at the frequency $\sim\omega_{\mathrm{pe}}$, but also in ensuring that the spatial distribution of the amplitude corresponds to the formfactor $E(x)$ from equation (B.1).

The author of paper [10] started by considering the splitting of hydrogenic spectral lines at the fixed value of x. Then he averaged the result over the formfactor $E(x)$ from equation (B.1) to produce new results.

Blochinzew in his pioneering work of 1933 [11] considered the splitting of a model hydrogen line, consisting of just one Stark component, under a linearly-polarized electric field $\mathbf{E}_0 \cos \omega t$. He demonstrated that the model line splits up in satellites separated by $p\omega$ ($p = \pm 1, \pm 2, \pm 3, \ldots$) from the unperturbed frequency ω_0 of the spectral line:

---

[1] It should be noted that paper [8] by Stamm *et al* had a broad title 'Line shapes in turbulent plasmas'. Regrettably, the authors of paper [8] seem to be ignorant of dozens and dozens of theoretical and experimental papers on the subject of line shapes in turbulent plasmas that started in Sholin's group as early as in the 1970s and continued by various theoretical and experimental groups around the world through 2017—see, e.g. paper [9], book [2] and references from paper [9] and book [2].

$$S(\Delta\omega/\omega) = \sum_{p=-\infty}^{+\infty} \left[J_p^2(X_k\varepsilon)\right](\delta(\Delta\omega/\omega) - p), \qquad (B.4)$$

where $J_p(u)$ are the Bessel functions, ε is the scaled amplitude of the field:

$$\varepsilon = 3\hbar E_0/(2Z_r m_e e\omega). \qquad (B.5)$$

Here Z_r is the nuclear charge of the radiating atom/ion and

$$X_k = nq - n_0 q_0, \qquad (B.6)$$

where n, q and n_0, q_0 are the principal and electric quantum numbers of the upper and lower energy levels, respectively, involved in the radiative transition. From the physical point of view, the greater the product $X_k\varepsilon$, the larger is the phase modulation of the atomic oscillator. In paper [12] Blochinzew's result was generalized to profiles of real, multicomponent hydrogenic spectral lines in the 'reduced frequency' scale as follows (presented later also in book [4], section 3.1)

$$S(\Delta\omega/\omega) = \sum_{p=-\infty}^{+\infty} I(p,\ \varepsilon)\delta(\Delta\omega/\omega) - p, \qquad (B.7)$$

where

$$I(p,\ \varepsilon) = \left[f_0\delta_{p0} + 2\sum_{k=1}^{k_{\max}} f_k J_p^2(X_k\varepsilon)\right] \bigg/ \left(f_0 + 2\Sigma f_k\right). \qquad (B.8)$$

Here f_0 is the total intensity of all central Stark components, f_k is the intensity of the lateral Stark component with the number $k = 1, 2, \ldots, k_{\max}$.

In the typical case of the strong modulation ($X_k\varepsilon \gg 1$), from equations (B.4), (B.7), (B.8) it is seen that there could be numerous satellites of significant intensities. Often the situation is such that the individual satellites merge together by broadening mechanisms. In this situation only the envelope of these satellites can be observed. The most intense part of the satellites envelope has the shape of the Airy function, as shown in paper [12] and reproduced in book [4], section 3.1. Based on these analytical results, the following practical formulas have been derived and presented in paper [12] and reproduced in book [3] (section 3.1) for the position $p_{\max}$ of the satellite having the maximum intensity (and thus corresponding to the experimental peak)

$$p_{\max}(a) = a + (a/2)^{1/3}d_{\mathrm{Ai}} = a - 0.809(a)^{1/3}, \qquad a = X_k\varepsilon, \qquad (B.9)$$

where $d_{\mathrm{Ai}} = -1.019$ is the first zero of the derivative of the Airy function.

In paper [10] the author started the averaging over the formfactor $E(x)$ from equation (B.1) by substituting $E(x)$ into the argument of the Bessel function in equation (B.4) and integrating over x from $-\lambda$ to λ, or equivalently

$$f(p, a, b) = (1/b) \int\int_{-b/2}^{b/2} dy J_p^2[a/ch(y)], \qquad (B.10)$$

where he denoted

$$b = L/\lambda, \quad y = x/\lambda. \qquad (B.11)$$

Since the maximum intensity has the satellite at the position $p_{\max}(a)$ given by equation (B.9), then its spatially integrated intensity from equation (B.10) is $I = f$ $[p_{\max}(a), a, b]$. Figure B1 shows I versus b for $a = 10$ (solid line), $a = 15$ (dashed line), and $a = 20$ (dotted line). It is seen that the integrated intensity of the most prominent satellite decreases as either a increases (e.g. the field amplitude E_0 increases) or as b increases (i.e. the distance L between Langmuir solitons in the sequence increases).

Figure B2 presents a three-dimensional plot of the spatially integrated intensity of the most prominent satellite versus both a and b.

Figure B3 shows a three-dimensional plot of the ratio $f[p_{\max}(a), a, b]/f[0, a, b]$ versus a and b. This is the ratio of the spatially integrated intensity of the most prominent satellite to the spatially integrated intensity of the 'zeroth' satellite, the latter being the intensity at the unperturbed position of the one-component spectral line. It is seen that this ratio is generally a non-monotonic function of the scaled amplitude a of the solitons electric field.

Then the author of paper [10] proceeded to real, multi-component hydrogenic spectral lines. As an example, he utilized the Ly-beta line in the observation perpendicular to the solitons electric field.

Figure B4 presents the profile of the Ly-beta line versus the scaled distance $\Delta\omega/\omega$ from the unperturbed position of this line for the (differently) scaled amplitude $\varepsilon = 3\hbar E_0/(2Z_r m_e e\omega) = 1$ and the scaled distance $b = L/\lambda = 2$ between the Langmuir solitons in the sequence (solid line). Also shown is the corresponding profile for the case of the non-solitonic Langmuir waves for the same value of $\varepsilon = 1$ (dashed line).

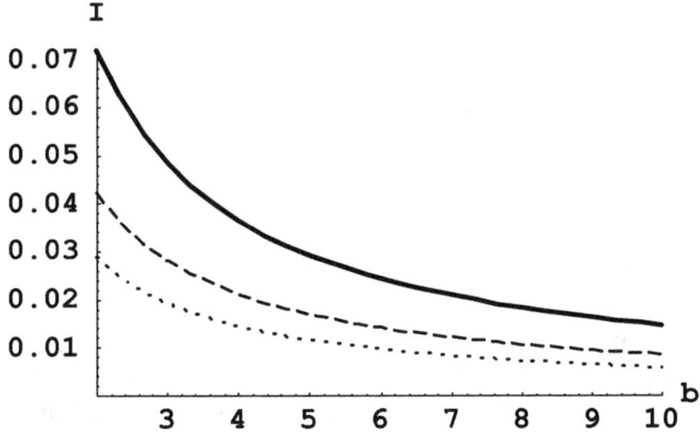

Figure B1. Spatially integrated intensity I of the most prominent satellite versus the scaled distance $b = L/\lambda$ of Langmuir solitons in the sequence for three values of the scaled amplitude $a = 3\hbar X_k E_0/(2Z_r m_e e\omega)$ of the solitons electric field: $a = 10$ (solid line), $a = 15$ (dashed line), and $a = 20$ (dotted line) [10].

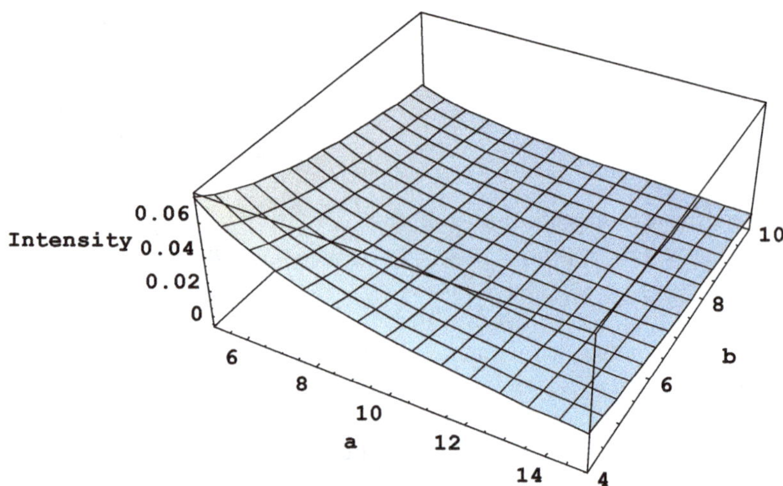

Figure B2. Spatially integrated intensity of the most prominent satellite versus the scaled amplitude $a = 3\hbar X_k E_0/(2Z_r m_e e\omega)$ of the solitons electric field and versus the scaled distance $b = L/\lambda$ of Langmuir solitons in the sequence [10].

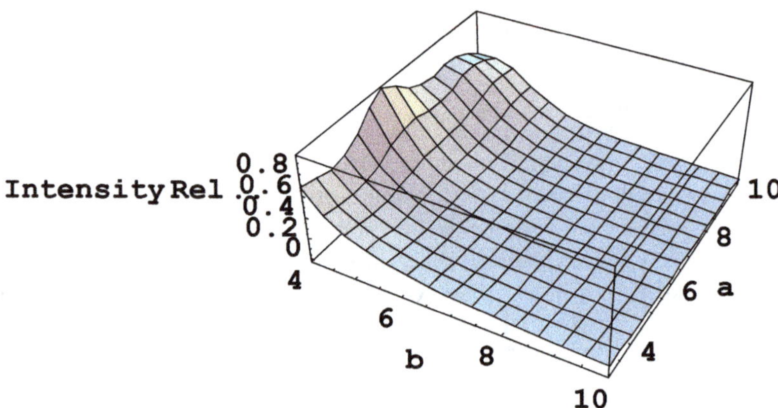

Figure B3. The ratio $f[p_{max}(a), a, b]/f[0, a, b]$ of the spatially integrated intensity of the most prominent satellite to the spatially integrated intensity of the 'zeroth' satellite versus the scaled amplitude $a = 3\hbar X_k E_0/(2Z_r m_e e\omega)$ of the solitons electric field and versus the scaled distance $b = L/\lambda$ of Langmuir solitons in the sequence [10].

The direction of the observation is perpendicular to vector $\mathbf{E}_0$. The profiles are continuous (rather than being a set of satellites isolated from each other) because additional broadening mechanisms (the Stark broadening by plasma microfields and the Doppler broadening) were taken into account in the amount of $\delta\omega = 2\omega$. The latter relation could be satisfied, for example, in plasmas of multi-charged ions produced by a powerful Nd-glass laser, where at the surface of the critical density, the electron density is $N_e = 10^{21}$ cm^{-3} (or slightly higher due to relativistic effects) and the temperature would be up to $T \sim 10^3$ eV.

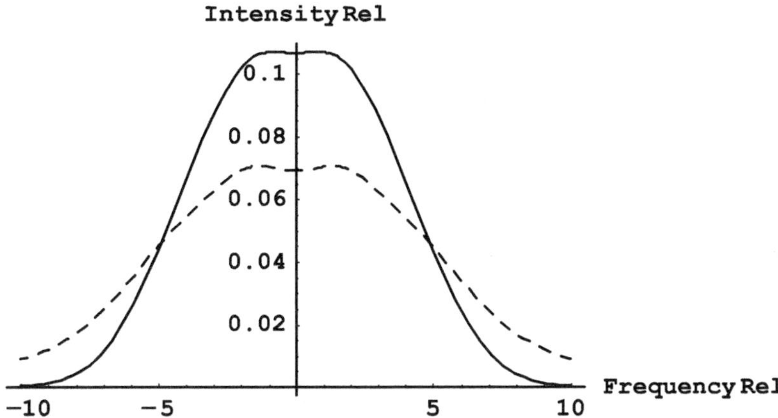

Figure B4. The profile of the Ly-beta line versus the scaled distance $\Delta\omega/\omega$ from the unperturbed position of this line for the (differently) scaled amplitude $\varepsilon = 3\hbar E_0/(2Z_r m_e e\omega) = 1$ and the scaled distance $b = L/\lambda = 2$ of Langmuir solitons in the sequence (solid line) [10]. Also shown is the corresponding profile for the case of the non-solitonic Langmuir waves for the same value of $\varepsilon = 1$ (dashed line). The direction of the observation is perpendicular to vector $\mathbf{E}_0$.

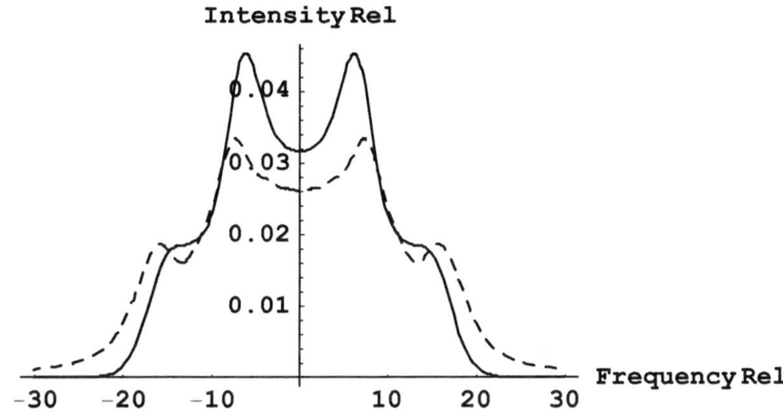

Figure B5. The same as figure B4, but for stronger Langmuir waves, corresponding to $\varepsilon = 3$ [10].

It is seen that in the case of the solitons, the profile is narrower than in the non-solitonic case. It is also seen that both profiles have practically the bell-shape without any significant features.

Figure B5 shows the same as figure B4, but for stronger Langmuir waves, corresponding to $\varepsilon = 3$. Both profiles have features. Namely, in the non-solitonic case (dashed line), the profile has two maxima both in the red and blue sides. In the case of solitons, the primary maximum in each side remains, but the secondary maximum in each side transforms into a shoulder.

Figure B6 shows the same as figure B5, but for even stronger Langmuir waves, corresponding to $\varepsilon = 6$. Both profiles have more features than in figure B5: three maxima both in the red and blue sides. In the case of solitons, the second maximum

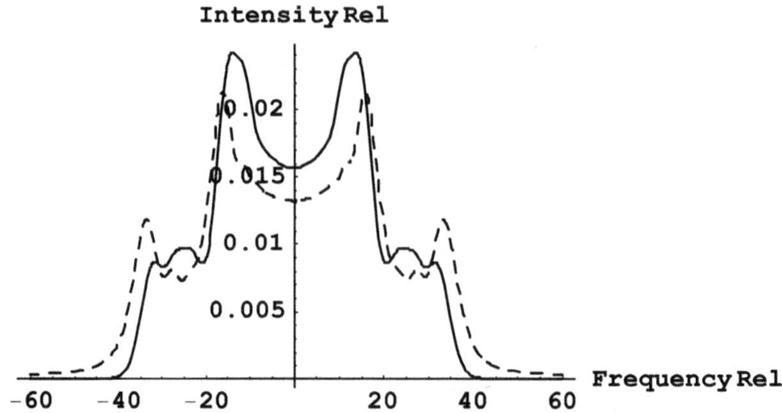

Figure B6. The same as figure B5, but for even stronger Langmuir waves, corresponding to $\varepsilon = 6$ [10].

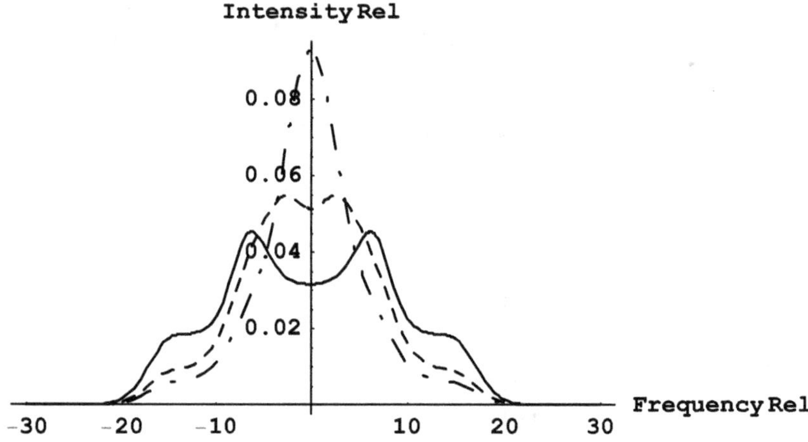

Figure B7. Dependence of the Ly-beta profiles for the case of the Langmuir solitons on the scaled distance $b = L/\lambda$ between the Langmuir solitons in the sequence: for $b = 2$ (solid line), $b = 4$ (dashed line), and $b = 6$ (dash-dotted line) [10].

in each side is more pronounced than for the non-solitonic case. For the third maximum the situation is opposite: the case of solitons, the third maximum is less pronounced than for the non-solitonic case.

Then the author of paper [10] exhibited how the Ly-beta profiles for the case of the Langmuir solitons, depend on the scaled distance $b = L/\lambda$ between Langmuir solitons in the sequence. In figure B7 the Ly-beta profiles corresponding to $\varepsilon = 3$, are presented for $b = 2$ (solid line), $b = 4$ (dashed line), and $b = 6$ (dash-dotted line). It is seen that as the scaled distance $b = L/\lambda$ between Langmuir solitons in the sequence increases, the features (such as maxima, minima, shoulders) gradually disappear.

The above shows that the diagnostic of Langmuir solitons, while employing, for example, the Ly-beta line, can be based on the following feature. In the non-solitonic case, there could be distinct secondary maxima in each wing of the line, whereas in the case of solitons the would-be secondary maxima look more or less like shoulders —see figures B5 and B6, and the solid line in figure B7.

In summary, in paper [10] the author suggested using the following manifestations of Langmuir solitons as the diagnostic tool. For the case of the Langmuir solitons, some maxima in the line profiles become less pronounced or even transform into shoulders—compared to the non-solitonic Langmuir waves of the same amplitude. Also, as the amplitude of the Langmuir solitons increases, more features (such as maxima and minima) appear in the line profiles. However, when the separation between the solitons within their set increases, there are less features in the line profiles.

These manifestations can be used for the experimental determination of the amplitude of the Langmuir solitons and of their separation from each other in their sequence.

References

[1] Kadomtsev B B 1982 *Collective Phenomena in Plasma* (Oxford: Pergamon)

[2] Oks E 2017 *Diagnostics of Laboratory and Astrophysical Plasmas Using Spectral Lines of One-, Two-, and Three-Electron Systems* (Singapore: World Scientific)

[3] Griem H R 1997 *Principles of Plasma Spectroscopy* (Cambridge: Cambridge University Press)

[4] Oks E 1995 *Plasma Spectroscopy: The Influence of Microwave and Laser Fields* (Berlin: Springer)

[5] Oks E 2017 *J. Phys. Conf. Ser.* **810** 012006

[6] Hannachi I, Stamm R, Rosato J and Marandet Y 2016 *Europhys. Lett.* **114** 23002

[7] Hannachi I, Meireni M, Rosato J, Stamm R and Marandet Y 2018 *Contrib. Plasma Phys.* **58** 583

[8] Stamm R *et al* 2017 *Eur. Phys. J.* D **71** 68

[9] Oks E *et al* 2017 *Opt. Express* **25** 1958

[10] Oks E 2019 *Atoms* **7** 25

[11] Blochinzew D I 1933 *Phys. Z. Sov. Union* **4** 501

[12] Oks E 1984 *Sov. Phys. Doklady* **29** 224

IOP Publishing

Eugene Oks

Appendix C

Profiles of hydrogenic spectral lines under stochastic electric fields of plasma turbulence: applications to diagnostics of the Langmuir turbulence

At sufficiently high electron densities $N_e \gg 10^{18}$ cm^{-3}, the Langmuir turbulence can cease to be quasimonochromatic and rather represents a broadband electric field with the peak intensity at the plasma electron frequency ω_{pe}, where

$$\omega_{\mathrm{pe}} = (4\pi e^2 N_e/m_e)^{1/2} \tag{C.1}$$

(here e and m_e are the electron charge and mass, respectively). This happens mostly due to electron collisions. The electric field of the Langmuir turbulence becomes multi-mode and can be even stochastic.

In this scenario the Langmuir turbulence can lead to a broadening of hydrogenic spectral lines. In the 1970s, this effect was pointed out by Sholin [1] and then developed in more detail by Oks and Sholin [2]. This effect was used for interpreting some experimental results in papers [3–5].

A further theoretical study related to this physical situation was performed by Gavrilenko [6] in 1996. Specifically, he considered modifications of profiles of hydrogen spectral lines under a multi-mode *non-monochromatic* linearly-polarized electric field[1].

In paper [6] Gavrilenko obtained his results for the case where the *power spectrum* of the stochastic electric field is *Lorentzian*. In paper [8] the author extended Gavrilenko's results to the scenario where the *power spectrum* of the stochastic

---

[1] It should be mentioned that in 1958, Lifshitz [7] studied theoretically the influence of a multi-mode *monochromatic* linearly-polarized electric field on a model hydrogen line, consisting of just one Stark component. That study was not relevant to the problem of broadening hydrogenic spectral lines by the multi-mode *non-monochromatic* electric field of the Langmuir turbulence in plasmas of sufficiently high electron densities.

doi:10.1088/978-0-7503-3375-7ch9   © IOP Publishing Ltd 2020

electric field is *Gaussian*. In addition, he studied theoretically the general case of hydrogenlike spectral lines—rather than only hydrogen spectral lines considered by Gavrilenko [6].

In paper [8] the author first deduced a general analytical result for the correlation function, whose Fourier transform determines the shape of the spectral line. Then he demonstrated that when the power spectrum of the field is sufficiently broad, hydrogenic line profiles are significantly narrower in his case compared to Gavrilenko's case—despite the power spectra of the field in both cases having the same full width at half maximum (FWHM). Below are a few more details.

Following notation by Gavrilenko [6], the correlation function of the field can be represented in the form

$$\{\mathbf{E}(t)\mathbf{E}(t + \tau)\}_{\mathrm{av}} = BG(\tau), \qquad B = \{\mathbf{E}^2\}_{\mathrm{av}} \tag{C.2}$$

where $G(\tau)$ is a correlation coefficient.

In paper [8] the author considered the case where the correlation coefficient is

$$G(\tau) = \exp(-\tau^2/g^2)\cos \omega\tau, \tag{C.3}$$

so that the *power spectrum* of the field $\mathbf{E}(t)$ has the *Gaussian* form. The author of paper [8] introduced the following notation

$$b = C_{\alpha\beta}^2 B, \tag{C.4}$$

where

$$C_{\alpha\beta} = 3(n_a q_\alpha - n_b q_\beta)/(2Z). \tag{C.5}$$

In equation (C.5), Z is the nuclear charge; n and q are the principal and electric quantum numbers, respectively, of the upper (a, α) and lower (b, β) Stark sublevels involved in the radiative transition ($q = n_1 - n_2$, where n_1 and n_2 are the parabolic quantum numbers). In equations (C.4), (C.5), and below, the atomic units are used:

$$\hbar = e = m_e = 1.$$

Further, the author of paper [8] denoted as D the detuning in the frequency scale from the unperturbed position of the spectral line:

$$D = \Delta\Omega - \delta_{ab}. \tag{C.6}$$

Below are three figures from paper [8] as the illustrations of the analytical results from paper [8] for the correlation coefficient of the electric field given by equation (C.3). In all figures below, the quantities b, g, and D are measured in units of the carrier frequency ω (for example, $D = (\Delta\Omega - \delta_{ab})/\omega$).

Figure C1 exhibits the profile of any Stark component for $b = 20$ and $g = 0.2$ (bold solid line). Also shown is the Lorentzian of $\mathrm{FWHM_L} = \pi^{1/2}bg$ (dashed line) and the Gaussian of $\mathrm{FWHM_G} = 2(2b \ln 2)^{1/2}$ (thin solid line). It is seen that the bulk of the profile is close to the Lorentzian shape and that in the wings there occurs the transition to the Gaussian shape. This kind of spectral line shape is a counterintuitive result.

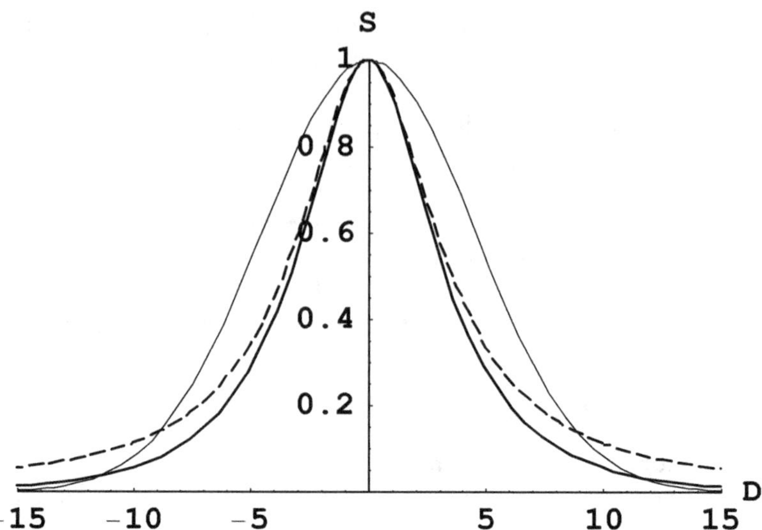

Figure C1. The calculated profile of any Stark component for $b = 20$ and $g = 0.2$ (bold solid line) versus the scaled dimensionless detuning $D = (\Delta\Omega - \delta_{ab})/\omega$. Also shown is the Lorentzian of $\text{FWHM}_\text{L} = \pi^{1/2}bg$ (dashed line) and the Gaussian of $\text{FWHM}_\text{G} = 2(2b \ln 2)^{1/2}$ (thin solid line) [8]. The profiles are peak-normalized. The quantities g and b from equations (C.3) and (C.4) are in units of the carrier frequency ω. Reprinted from [8], copyright 2020, with permission from Elsevier.

In figure C2, the calculated three-dimensional plot demonstrates the transformation of the profile of any Stark component as the quantity g varies, while the quantity $b = 30$. It is seen that as the quantity g increases, so does the width of the profile

In figure C3 the calculated three-dimensional plot demonstrates the transformation of the profile of any Stark component as the quantity b varies, while the quantity $g = 0.2$. It is seen that as the quantity b increases, so does the width of the profile.

From figures C2 and C3 one can clearly see that as b or g increases, so does the width of the profile of any Stark component. This is consistent with the analytical results from paper [8].

By analyzing the corresponding theoretical profiles of the hydrogenic Ly-beta line, the author of paper [8] proposed a *new diagnostic method allowing for the first time not only to measure experimentally the average field* of the Langmuir turbulence in dense plasmas, but also to find out *the information on the power spectrum* of the Langmuir turbulence. This diagnostic method is not limited to using the Ly-beta line of hydrogenic atoms/ions. This method would work while using other intense hydrogenic spectral lines that, like the Ly-beta line, do not have the central Stark components. Examples are the hydrogenic spectral lines Ly-delta, Balmer-beta, and Balmer-delta. Thus, this method could serve as a tool for the experimental testing of the field correlation function.

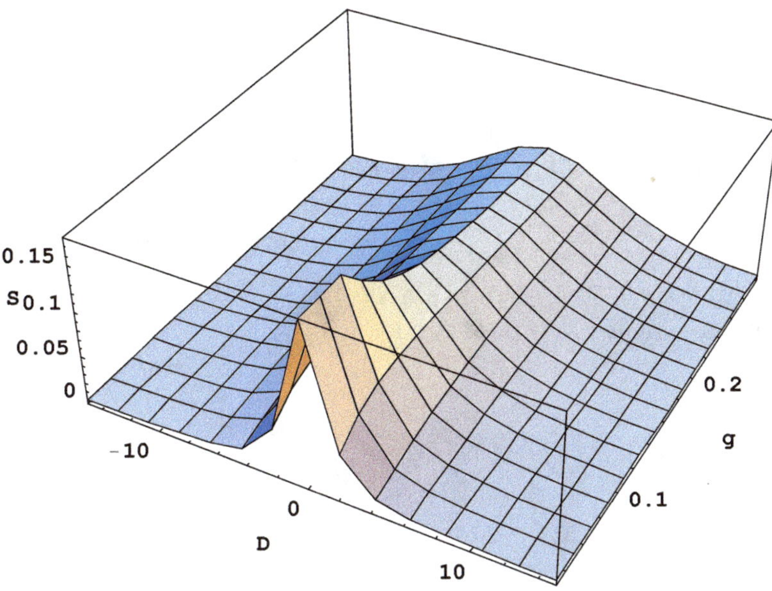

Figure C2. The calculated three-dimensional plot showing the transformation of the profile of any Stark component as the quantity g varies, while the quantity $b = 30$ [8]. The quantities g and b from equations (C.3) and (C.4) are in units of the carrier frequency ω. Reprinted from [8], copyright 2020, with permission from Elsevier.

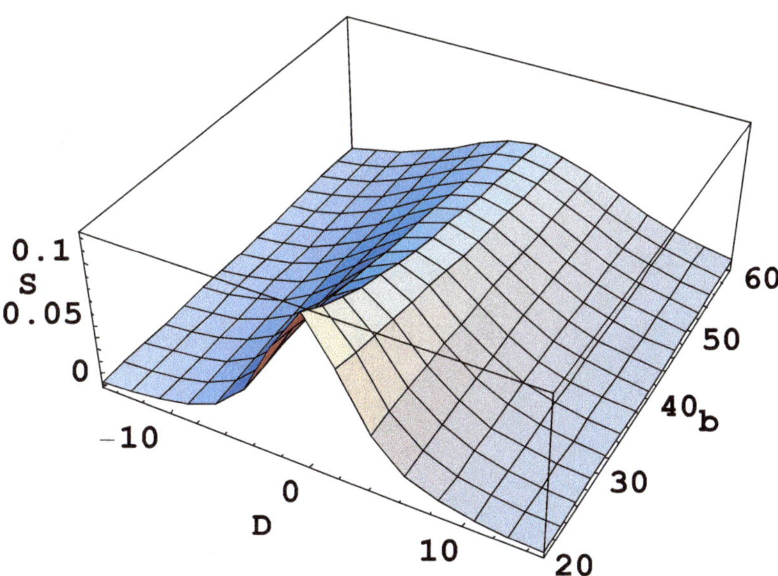

Figure C3. The calculated three-dimensional plot showing the transformation of the profile of any Stark component as the quantity b varies, while the quantity $g = 0.2$ [8]. The quantities g and b from equations (C.3) and (C.4) are in units of the carrier frequency ω. Reprinted from [8], copyright 2020, with permission from Elsevier.

References

[1] Sholin G V 1970 *Sov. Phys. Doklady* **15** 1040
[2] Oks E and Sholin G V 1975 *Sov. Phys. JETP* **41** 482
[3] Zakatov L P, Plakhov A G, Shapkin V V and Sholin G V 1971 *Sov. Phys. Doklady* **16** 451
[4] Karfidov D M and Lukina N A 1997 *Phys. Lett.* A **232** 443
[5] Oks E 2016 *J. Phys. B: At. Mol. Opt. Phys.* **49** 065701
[6] Gavrilenko V P 1996 *Pis'ma v Zh. Tech. Phys. (Sov. Phys. Tech. Phys. Lett.)* **22** 23 (in Russian)
[7] Lifshitz E V 1958 *Sov. Phys. JETP* **26** 570
[8] Oks E 2020 *Spectrochim. Acta* B **167** 105815